Building Sub-Contract
Documentation

Also of interest

Contract Documentation
for Contractors
Second Edition
Vincent Powell-Smith & John Sims
0–632–02275–2

Standard Letters for Building Contractors
Second Edition
David Chappell
0–632–03452–1

The JCT Design and Build Contract
David Chappell & Vincent Powell-Smith
0–632–02081–4

In preparation

Building Contract Claims
Third Edition
Vincent Powell-Smith & John Sims
0–632–03646–X

Building Sub-contracts
A Guide for Specialist Contractors
Silviu Klein
0–632–03763–6

Building Sub-Contract Documentation

David Chappell
BA (Hons Arch), MA (Arch), MA (Law), PhD, RIBA

and

Vincent Powell-Smith
LLB (Hons), LLM, DLitt, FCIArb, DipCom, DSLP, AIArbA

OXFORD

BLACKWELL SCIENTIFIC PUBLICATIONS

LONDON EDINBURGH BOSTON

MELBOURNE PARIS BERLIN VIENNA

© David Chappell & Ingramlight
Properties Ltd 1994

Blackwell Scientific Publications
Editorial Offices:
Osney Mead, Oxford OX2 0EL
25 John Street, London WC1N 2BL
23 Ainslie Place, Edinburgh EH3 6AJ
238 Main Street, Cambridge,
  Massachusetts 02142, USA
54 University Street, Carlton,
  Victoria 3053, Australia

Other Editorial Offices:
Librairie Arnette SA
1, rue de Lille
75007 Paris
France

Blackwell Wissenschafts-Verlag GmbH
Düsseldorfer Str. 38
D-10707 Berlin
Germany

Blackwell MZV
Feldgasse 13
A-1238 Wien
Austria

First published 1994

Set by DP Photosetting, Aylesbury, Bucks
Printed and bound in Great Britain by
Hartnolls Ltd, Bodmin, Cornwall

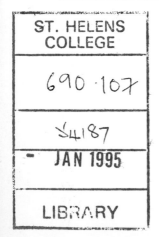
DISTRIBUTORS

Marston Book Services Ltd
PO Box 87
Oxford OX2 0DT
(*Orders:* Tel: 0865 791155
          Fax: 0865 791927
          Telex: 837515)

USA
Blackwell Scientific Publications, Inc.
238 Main Street
Cambridge, MA 02142
(*Orders:* Tel: 800 759-6102
               617 876 7000)

Canada
Oxford University Press
70 Wynford Drive
Don Mills
Ontario M3C 1J9
(*Orders:* Tel: (416) 441-2941)

Australia
Blackwell Scientific Publications Pty Ltd
54 University Street
Carlton, Victoria 3053
(*Orders:* Tel: 03 347-5552)

British Library
Cataloguing in Publication Data
A Catalogue record for this book is available
from the British Library

ISBN 0–632–02084–9

Library of Congress
Cataloging in Publication Data
Chappell, David.
    Building sub-contract documentation/
    David Chappell and Vincent Powell-Smith.
        p.   cm.
    Includes index.
    ISBN 0–632–02084–9
    1. Construction contracts—Great
Britain.   2. Construction industry—
Great Britain—Subcontracting.
I. Powell-Smith, Vincent.
II. Title
KD1641.D478   1994
343.73′07869—dc20
[347.3037869]   93-45871
                                            CIP

Contents

Preface

Sub-contractors account for the bulk of the work actually done in the building industry and they seem to generate more than their fair share of legal problems. Books especially written for the benefit of sub-contractors are relatively rare. In view of the success of our other books giving advice and sample letters for architects and for contractors, we thought it would be helpful to offer a similar service to sub-contractors.

Sub-contractors, in particular, find little time for letter writing and filling in forms other than the absolutely vital statutory forms. The smallest tend to operate quite informally. Many quotations are given on the telephone and some contracts are still concluded with a handshake. Sub-contractors build up relationships with certain contractors which are seen to need careful fostering; the constant writing of letters and the sending of contractual notices is not looked upon as conducive to good relations.

These commercial realities must always take precedence and the following advice and sample letters take these factors into account while not sacrificing contractual principles. Precise circumstances, however, must always dictate when a letter is sent and its precise tone. Therefore, we have made our letters formal and fairly neutral in tone. It is an imprudent sub-contractor who resists sending an important letter for fear that the contractor will not employ him again. Failure to send contractual notices at the right time can result in loss of financial entitlements and exposure to claims and contra-charges. At the other end of the scale, there are some very large sub-contractors. We hope that they too will benefit from our work.

We have aimed to write a simple, easily referenced book in which the busiest sub-contractor can find the help he is most likely to need to deal with common problems. Standard sub-contracts DOM/1, DOM/2, NAM/SC, NSC/C and associated standard documents are covered and much of the content is of general application. We provide a brief commentary explaining the implication of each document and highlighting the points to be watched. The advice is given under topic areas: extensions of time, payment, set-off, etc.

Through the publishers, we welcome comments and instances of further problems for inclusion in future editions of this book.

Acknowledgement

We are grateful to RIBA Publications Ltd for permission to reproduce NAM/T and the attestation from NSC/A.

David Chappell   *Vincent Powell-Smith*
Chappell-Marshall Limited   University of Malaya
176 Easterly Road   59100 Kuala Lumpur
Leeds LS8 3AD   Malaysia
November 1993

Chapter 1

Introduction

1.01 General

The majority of the actual work on any major building project is carried out by sub-contractors. This is inevitable as buildings and structures become more complex and it is recognized by all the major forms of standard building contract, which contain provisions permitting the main contractor to sub-let parts of the work with the consent of the employer or the architect. In some cases sub-contracting is mandatory, because the sub-contractor is selected through the systems of 'nomination' or 'naming'.

Where the main contract is in one or other of the commonly used standard forms, such as those produced by the Joint Contracts Tribunal (JCT), supporting forms of sub-contract are published. The Minor Works form is deficient in this respect. It is essential that these forms are used, because the appropriate main contract obligations are stepped down in the sub-contract, which also gives the sub-contractor rights and remedies corresponding to those which the main contract gives the main contractor against the employer. The main and sub-contracts are complementary to each other and there is (or should be) no conflict between the two sets of conditions.

If the sub-contractor is not of the main contractor's own choice, but is imposed upon him by the employer, i.e. is 'nominated' or 'named' through the detailed provisions in the main contract, the use of standard form sub-contract documentation is often mandatory. This is the case, for example, under both the JCT Standard Form of Building Contract, 1980 edition (JCT 80) and the JCT Intermediate Form of Building Contract, 1984 (IFC 84). In contrast, under the Association of Consultant Architects' Form of Building Agreement, 2nd edition, 1984, revised 1990 (ACA 2), although there is a supporting standard sub-contract form, its use is not mandatory even if the sub-contractor is chosen by the architect as a 'named' sub-contractor.

Even where there is no mandatory documentation, it is essential that a recommended standard form of sub-contract be used if the sub-contractor's interests are to be safeguarded.

The perils of sub-contracting on non-standard or in-house terms are illustrated by the well-known case of *Martin Grant & Co Ltd* v. *Sir Lindsay Parkinson & Co Ltd* (1984) 3 Con LR 12, where Grant were framework sub-contractors to Parkinson on several local authority housing projects. The main contracts were in JCT 63 form. The sub-contracts, entered into in 1971, were in no-standard form and the fluctuation clause in it was of a limited kind. The main contracts contemplated that the work would be completed within three years, but there were substantial delays in the performance of the main contracts. Grant's sub-contract went on for no less than five years. As a result they incurred substantial losses for which they were not reimbursed.

The Court of Appeal refused to extricate them from the difficulty by implying a term that the main contractor would make sufficient work available to the sub-contractors to enable them to maintain reasonable progress and to execute their work in an efficient and economic manner.

The sub-contract required Grant to execute their work 'at such time or times and in such manner as the [main] contractor shall direct or require'. The sub-contract meant, in effect, that if the period of the main contract was extended then the sub-contract period would be extended also, but that during the whole of the sub-contract period Grant would do such portions of the work at such times as might be required by the main contractor. In other words, there was a clear risk for Grant that the main and sub-contracts might go on much longer than was originally contemplated as, indeed, proved to be the case.

Some main contractors are notorious for producing in-house forms of contract which sometimes bear close typographical resemblance to the standard sub-contract forms. Such forms usually contain innovatory conditions in favour of the main contractor (such as 'pay-when-paid' clauses) and omit important protections for the sub-contractor. This is commonly the case wiith domestic sub-contractors under JCT 80.

If the main contract is on JCT 80 terms, the only terms for any domestic sub-contract should be Domestic Sub-Contract Conditions (DOM/1) which is approved by the Building Employers Confederation, the Federation of Associations of Specialists and Subcontractors and the Committee of Associations of Specialist Engineering Contractors, which are the representative bodies of both main and sub-contractors.

There are several reasons for this. In the first place, it is a document agreed between organisations representing both parties and is, therefore, framed to protect both their interests. This is exemplified by the clauses which give each party rights against the other to recover any direct loss and/or expense arising from specified defaults of the other. Secondly, DOM/1 is so drafted as to be compatible with JCT 80 and to step down the relevant main contract terms. Thirdly, being a negotiated form of sub-contract agreed by the industry's representatives, it will not be interpreted *contra proferentem*. This is a principle or rule of contract interpretation that any ambiguities in a document which all other methods of interpretation have failed to resolve will be interpreted against the person seeking to rely on it. Finally, an in-house form might well be regarded as the contractor's 'standard terms of business' for the purposes of the Unfair Contract Terms Act 1977.

These last two points are often overlooked by those main contractors who use sub-contract forms of their own devising in order to give themselves increased rights against the sub-contractor. A common example is where the main contractor's in-house form gives him an express right to set-off against payments due to the sub-contractor unquantified claims or even claims arising under other sub-contracts. One such in-house clause has the effect of making the main contractor's own estimate of amounts due binding and conclusive as to both liability and amount until final ascertainment. It reads:

'If the sub-contractor is in breach of any of his obligations [as to progress] he shall, without prejudice to and pending the final ascertainment or agreement between the parties as to the loss or damage suffered or incurred or which may be suffered or incurred by the contractor in consequence thereof, forthwith pay or allow to the contractor such sum as the contractor shall bona fide estimate as the amount of such loss or damage, and such estimate shall be binding and conclusive upon the sub-contractor until such final ascertainment or agreement.'

This type of clause has failed to withstand legal challenge in the courts. In contrast, under clause 23 of DOM/1, the main contractor's right of set-off is confined to claims which have been agreed or at least quantified with reasonable accuracy and a detailed procedure for the excercise of the right of set-off is laid down. The sub-contractor's rights are further protected by clause 24 which provides for rapid adjudication by an independent person if the set-off is disputed.

Regrettably, the past few years have seen an increase in the practice of main contractors seeking to impose unreasonable or even impossible terms on their sub-contractors. One example has already been given. Another is that one of us has recently seen a sub-contract which was superficially in DOM/1 form, but which had been heavily amended by the main contractor to remove many of the sub-contractor's safeguards. One of the particularly objectionable amendments sought to make the sub-contractor absolutely liable for any loss or damage to any of his materials howsoever caused up to the date of practical completion of the main contract works and, of course, even if they had been incorporated into the works. It is a sad fact that some main contractors of the highest probity in their general dealings seem to lose all sense of principle and morality in their dealings with sub-contractors.

Even where the industry's negotiated standard form sub-contracts are used, sub-contractors often lose out because of their own inadequate administration and a failure to appreciate the importance of the inter-relation of their own activities with those of the main contractor and other sub-contractors. A great many of the difficulties which arise in practice – and all too frequently end as disputes to be settled in arbitration or litigation – could have been avoided if the procedures prescribed in the relevant form of sub-contract had been followed. This applies equally to main contractors, of course, as the growing number of cases on the exercise of the main contractor's right of set-off under standard sub-contract forms illustrates.

At various points, the sub-contracts lay down procedures which must be followed – building sub-contracts are as much procedural as legal in their effect and all too often they do not clearly differentiate between good practice and legal requirements. Sub-contracts also require that certain things be done at a particular time, e.g. the giving of notice or the making of an application in a claims situation. Failure to do what the sub-contract requires may result in loss of entitlement.

In many instances, good practice requires that the sub-contractor should set down the situation in writing, either because the sub-contract requires it or because prudence and business sense dictate this course. Sometimes the correct procedure will be spelled out in the sub-contract and many common difficulties can be resolved by a careful reading of the printed terms. In our experience, many contracting parties do not read the terms either before or after signing them – possibly because of the somewhat verbose language. This may be the fault of the draughtsman, but failure to both read and understand what the sub-contract says can be costly.

In the chapters which follow, we have endeavoured to cover some of the most common situations encountered in sub-contracting. We have not written a legal textbook, nor yet a complete procedural handbook. This volume is intended to be the sub-contractor's equivalent of *Contractual Correspondence for Architects*, 2nd edition, by David Chappell (1989, Legal Studies & Services (Publishing) Ltd) and *Contract Documentation for Contractors*, 2nd edition, by Vincent Powell-Smith and John Sims (1990, Blackwell Scientific Publications) which have evidently been been found useful by architects and main contractors. Our aim is to give practical guidance to sub-contractors and their site and administrative staff.

We do not give examples of obvious letters or notices required by the sub-contract or desirable in particular circumstances. We have concentrated on situations which we know commonly give rise to difficulty and in particular those when things are likely to go wrong or where, because the situation is so obvious, it gets overlooked. We, therefore, provide a straightforward commentary and then suggest a possible wording for some notice or letter. We also reproduce some of the most commonly used standard documents. It must be emphasized that this is not a book of standard letters and documents to be used without modification or thought. The letters are merely examples and should be freely adapted. The reader cannot go far wrong if, when issuing a notice or drafting a letter, he or she follows the wording of the relevant sub-contract clause. In every case, reference should be made to the clause relied on.

The remainder of this book is written around the four most commonly-used standard forms of building sub-contract, all designed for use with one of the forms of main contract issued by the prolific Joint Contracts Tribunal. The sub-contracts are:

- *Domestic Sub-Contract DOM/1* which is for use with the JCT Standard Form of Building Contract (JCT 80) as the main contract. A domestic sub-contractor is any person or firm to whom the main contractor sub-lets any portion of the works with the architect's consent and who is not a nominated sub-contractor.
- *Domestic Sub-Contract DOM/2* is for use with the JCT Standard Form of Building Contract With Contractor's Design (CD 81) as the main contract. The conditions of sub-contract are based on DOM/1 conditions, but there are amendments designed to step down the special conditions of main contract associated with contractor design, reflecting the fact that there is no architect. In all those places in the main contract where one would normally expect to find the word 'architect', the 'employer' or 'employer's agent' occur instead.

- *Intermediate Sub-Contract NAM/SC* where the sub-contractor is 'named' under the JCT Intermediate Form of Contract (IFC 84). This document is supported by the JCT Form of Tender and Agreement NAM/T.
- *Nominated Sub-Contract NSC/C* is also for use with JCT 80, but only where the sub-contractor is 'nominated' by the architect under clause 35 and the nomination has been made involving completion of the standard forms NSC/T, NSC/W and NSC/A.

1.02   Tabular summaries

There are many notices and letters which must or should be given or sent by the sub-contractor under the four forms. The following is a tabular summary of the main notices and other documents whose issue is mandatory and not permissive.

Table 1   Mandatory notices, etc., from sub-contractor.

Content	DOM/1 and DOM/2 Clause No	NSC/A Clause No	NAM/SC Clause No
Request to concur in the appointment of an arbitrator	Article 3.1	Article 4	Article 4
Notice of restriction, limitation or exclusion	—	1.7.2	—
Notice specifying discrepancy or divergence in documents	2.3	1.8	—
Objection to compliance with instruction requiring change in obligation or restrictions	4.3	3.3.2	5.3
Statement of measures to establish no failure in repetitive work	—	—	5.6.1
Objection to compliance with instruction	—	—	5.6.4
Confirmation of directions issued otherwise than in writing	4.4	3.3.3	—
Request for authority for instruction	—	3.11	—
Loss or damage by specified perils, etc., where employer responsible for head contract insurance	8.2.2	6C.3.1	9.2.2
Production of evidence of insurance	9.2	6.7	9.2.2
Notice of delay	11.2.1	2.2.1	12.2
Particulars of expected effects of delay	11.2.2.1	2.2.2.1	—
Estimate of expected delay in completion	11.2.2.2	2.2.2.2	—
Further particulars, notices and estimates	11.2.2.3	2.2.2.3	—

Table 1   *Continues overleaf*

Table 1 *Contd.*

Content	DOM/1 and DOM/2 Clause No	NSC/A Clause No	NAM/SC Clause No
Application for direct loss and/or expense	13.1	4.38.1	14.1
Information in support of application	13.1.2	4.38.1.2	—
Details of loss and/or expense	13.3	4.38.1.3	—
Notice of disturbance of progress	—	4.39	—
Notice of practical completion	14.1	2.10	15.1
Daywork vouchers	16.3.4. & 17.3.3	4.6.4 & 4.13.3	—
Assessment of values for VAT purposes	19A.6.1 & 19B.6.1	5A.6.1 & 5B.6.1	17A.6.1 & 17B.6.1
Notice of withdrawal of consent	19B.5.2	5B.8.2	19B.5.2
Production of current tax certificate	20A.2.1	5C.2.1	18A.2.1
Notification of change in nominated bank account	20A.2.3	5C.2.3	18A.2.3
Notification of cancellation of tax certificate	20A.3.1	5C.3.1	18A.3.1
Notification of value of materials	20B.1.1	5D.1.1	18B.1.1
Notification of change in tax status	20B.5	5D.5	18B.5
Details in support of valuation statement	21.4.4	—	—
Request for application for interim payment	—	4.15.1	—
Representations as to interim payment	—	4.15.2	—
Supply of proof of discharge	—	4.16.1.1	—
Notice regarding late payment	21.6	4.21.1	19.6
Submission of documents necessary for adjustment of sub-contract sum/computation of final sub-contract sum	21.7.1 & 21.8.1	4.23.1.1 & 4.24.1	19.7.1
Statement of disagreement with set-off and particulars of counterclaim	24.1.1	4.30.1	22.1.1
Notice of arbitration	24.1.1	4.30.1.1	22.1.1
Request for action by adjudicator	24.1.2	4.30.1.2	22.1.1.2
Request for consent to assignment	26.1	3.13	24.1
Request for consent to sub-letting	26.2	3.14	24.2
Notice of default and determination	30.1	7.6	28.1
Notice of fluctuations	35.4.1 & 36.5.1	4A.4.1 & 4B.5.1	33.4.1
Evidence in respect of fluctuations	36.5.5	4B.5.5	33.4.5

Table 2 Mandatory notices, etc., from main contractor

Content	DOM/1 & DOM/2 Clause No	NSC/A Clause No	NAM/SC Clause No
Request to concur in the appointment of an arbitrator	Article 3.1	Article 4	Article 4
Issue copies of further drawings and details	—	—	2.3
Issue directions for correction of discrepancies	4.1.5.2	—	2.4
Notice requiring compliance with direction	4.5	3.10	5.4
Production of evidence of insurance	5.1	6.9	10.1
Inform architect of sub-contractor's notice of delay	—	2.2.1	—
Submission to architect of estimates and further notices	—	2.2.3	—
Grant of extension of time	11.3	2.3	12.2
Final extension of time	11.7	2.5	—
Notification of failure to complete on time	12.1	2.8	13
Pass notice of practical completion to	—	2.10	—
Statement of relevant events	13.2	—	—
Application to architect for interim payment	—	4.15.1	—
Passing representations as to interim payment to architect	—	4.15.2	—
Notification of discharge of certified payment	—	4.16.1.1	—
Supply of computation of sub-contract sum	—	4.23.2 & 4.24.2	9.8.2
Application for direct loss and/or expense	13.4	4.40	14.3
Information in support of application	13.4.2	—	—
Details of loss and/or expense	13.4.3	—	—
Notice of dissent to notice of practical completion	14.1	—	15.1
Notice of failure to supply VAT receipts	—	5A.10	17A.10
Document approved for VAT	19B.7.1	5B.7.1	17B.7.1
Reconciliation statement	19B.7.3	5B.7.3	17B.7.3
Withdrawal of approval	19B.8.1	5B.8.1	17B.8.1
Confirmation of satisfaction or non-satisfaction with tax certificate, etc.	20A.2.1	5C.2.1	18A.2.1
Notification of change of tax status	20A.3.2 & 20B.3	5C.3.2 & 5B.3	18A.3.2 & 18B.3

Table 2 *Continues overleaf*

Table 2 *Contd.*

Content	DOM/1 & DOM/2 Clause No	NSC/A Clause No	NAM/SC Clause No
Notification of requirement to make deduction	20A.5	5C.5	18A.5
Final payment	21.9.2	—	19.8.2
Notice of set-off	23.2.2	4.27.2	21.2.2
Statement of defence to counterclaim	24.2	4.31	22.2
Notice of default and determination	29.1	7.1	27.1
Notice of determination for insolvency	29.2	7.2	27.2
Request for assignment	29.4.2.1	7.4.2.1	—
Direction for removal of temporary buildings, etc.	29.4.3	7.4.2.3	27.3.1

Chapter 2

Tenders and Quotations

2.01 General

Named and nominated sub-contractors under sub-contracts NAM/SC and NSC/C respectively are normally appointed after a specific tendering process. NAM/SC procedures are dealt with in Chapter 3 and NSC/C is dealt with in Chapter 4.

Sub-contract tendering in other instances, if it exists at all, is generally a fairly informal business. It is common for a main contractor to telephone for a price and for the sub-contractor to receive the contractor's 'official order' months later after the main contract has been secured. It is sometimes said that there is only one real problem in sub-contract tendering: submitting a price which is high enough to cover costs and give a reasonable profit, but lower than the next tender. Sub-contractors approach the problem in different ways. Some appear to deal with enquiries in a state of wide-eyed innocence, barely glancing at the documentation and producing a price by what can only be described as empirical methods. Others subject every page of the enquiry to minute scrutiny, visit site, ask questions and produce the price after carefully weighing all the risk factors. That may be too broad a generalisation, but every contractor knows that the lowest tender is often submitted by a firm which has overlooked some crucial aspect of the work. It may appear that it is not worthwhile taking the time and trouble to submit a careful price, but there is no point in securing sub-contracts which turn out to be loss-makers.

Much has been said about the imposition of onerous conditions by main contractors in their sub-contract enquiries. What is sometimes forgotten is the way in which tenders are invited. In order to submit a price, the sub-contractor has to read all the information supplied in order to see what his obligations are and his rights and duties if his price is successful. The advantage of using standard sub-contract forms is that they usually contain provisions which distribute the risk in a fair and reasonable way. In addition, and very importantly, their terms properly reflect the main contract terms.

2.02   Conflicting terms and conditions

Difficulties arise if the main contractor wants to use his own form of sub-contract or if he insists on adding clauses to the contract. In either instance, conflicts will almost inevitably occur. A very common situation is that the sub-contractor is asked to tender on the main contractor's own sub-contract terms which conflict with the main contract and which are also in conflict with additional clauses listed on sheets of paper apparently added to the enquiry on a whim. To this situation may be added a letter of invitation including one or two further terms, a form of tender which is not quite applicable (for example, referring to bills of quantities when only a specification is provided) and the foundations for a future dispute are firmly in place.

Although it is fashionable to blame the main contractor in such cases and to impute doubtful motives, our experience is that most such conflicts are a result, not of scheming contractors, but of inexperienced people dealing with tender enquiries. A deliberately devious main contractor could reasonably be expected to make a better and more subtle job of things. Conflicting terms do not benefit either party and they should be dealt with as soon as they are discovered. Undoubtedly the best time to deal with them is before the sub-contract is executed. That is the case no matter what form of contract is being used (**document 2.02.1**). Conflicts are not limited to the use of non-standard forms. It is quite common to see standard forms used with one or more sheets of amending clauses or with accompanying documents which contain inconsistent terms. After execution, the conflict may have to be resolved by an arbitrator or a judge.

After the contract is executed, discrepancies are dealt with in DOM/1 and DOM/2 by clause 4.1.5, in NAM/SC by clause 2.4 and in NSC/C by clause 1.8. Each sub-contract has a different approach to the problem.

DOM/1 and DOM/2 state that the sub-contractor must immediately give written notice to the contractor specifying the discrepancy or divergence if found in or between two or more of:

- The documents in clause 2.3 of the main contract (JCT 80):
    The main contract drawings
    The main contract bills of quantity
    Architect's instruction (unless a variation under clause 13.2 of the main contract)
    Drawings or documents issued by the architect
    The numbered documents (These documents do not include the numbered documents which are part of the sub-contract DOM/1

or DOM/2, but do include all the numbered documents in all the nominated sub-contracts).
- Directions issued by the contractor (unless the direction requires a variation).
- The numbered documents which are part of the sub-contract.

Document 2.02.2 is appropriate.

It is now established that the wording of the clause does not place an obligation on the sub-contractor actively to search for discrepancies, merely to report them in writing, if found. Some contractors may try to take a hard line. Respond accordingly (**document 2.02.3**). In practice, it is difficult to imagine how the average domestic sub-contractor will be able to find discrepancies, because he will not have ready access to the documents listed in clause 2.3 of the main contract. On receipt of the sub-contractor's written notice, the contractor must issue a direction about the discrepancy. No particular time limit is stipulated, but probably it will be implied that the contractor's obligation is to issue the direction in such time that the regular progress of the work is not disturbed. If the contractor is lax, you must take a firm stance (**document 2.02.4**)

NAM/SC deals with inconsistencies (as it refers to them) very briefly. There is no obligation on the sub-contractor to notify the contractor if inconsistencies are found. Simply, the contractor must issue directions to correct inconsistencies in or between:

- The sub-contract documents (referred to in Article 1.1)
- Drawings or documents issued under clause 2.3 of the sub-contract NAM/SC

Whether or not the sub-contractor has an express duty to give written notice to the contractor, it is both prudent and a matter of common sense that he should notify the contractor if he finds any inconsistency (**document 2.02.5**). It is clear that once an inconsistency has been brought to the contractor's notice, he must act promptly to issue an appropriate direction.

NSC/C deals with discrepancies in clause 1.8 in somewhat similar manner to DOM/1, but with differences to reflect the nominated status of the sub-contractor. If the sub-contractor finds a discrepancy in or between the documents referred to in clause 2.3 of the main contract (see above), he must give the contractor a written notice specifying the discrepancy (**document 2.0.2.6**) and the contractor must forthwith send the notice to the architect requesting instructions.

DOM/1 and DOM/2 make no special provision for dealing with

errors or misdescriptions in bills of quantities. NAM/SC and NSC/C in clauses 3 and 1.12 respectively state that where bills of quantities are part of the sub-contract documents, the quality and quantity of the work shall be deemed to be as set out in the bills. The bills are to have been prepared in accordance with the *Standard Method of Measurement of Building Works*, 7th edition, unless expressly stated otherwise in respect of any specified item. NAM/SC merely provides that if there is any departure from the method the contractor must issue directions confirming a correcting instruction of the architect. NSC/C, not only provides for correction of departures, but also for misdescriptions, errors in quantity or omissions. There appears to be no requirement for the architect to issue an instruction in such circumstances, correction is to be made and it will be treated as if it was an architect's instruction requiring a variation. The wise sub-contractor will confirm the situation to the contractor (**document 2.02.7**). Although not expressly stated, it is thought that the same provision would be implied into NAM/SC, because the contract would not work otherwise.

2.02.1   Letter, if conflicting terms discovered before the sub-contract is executed

This letter is only suitable for use with DOM/1 or DOM/2
By fax and post

To the Contractor

Dear Sirs

[*Heading*]

Thank you for your letter of the [*insert date*] with which you
enclosed the sub-contract documents for our signature. We confirm
our telephone message earlier today that some of the terms appear
to conflict and we should welcome the opportunity to discuss and
resolve the conflicts before the contract documents are executed.
In view of the proposed start on site on the [*insert date*] we
suggest that the matter be treated with urgency. For your
information, the terms in conflict are: [*list the clause numbers, etc*]

[*If appropriate, add:*]

We also note that the following terms have been introduced since
we submitted our quotation/tender [*delete as appropriate*] figure.
These terms must be deleted.

[*Then:*]

We look forward to hearing proposed alternative meeting dates and
confirm that we are not prepared to commence work on site until
the contract terms are agreed.

Yours faithfully

2.02.2 Letter, if discrepancy found

This letter is only suitable for use with DOM/1 or DOM/2

To the Contractor

Dear Sirs

[*Heading*]

In accordance with clause 4.1.5 of the conditions of sub-contract, we give notice that we have discovered a discrepancy in/between [*delete as appropriate and specify the discrepancy, giving precise details of drawing numbers, numbered document references, architect's instructions and contractor's directions*]

In order to avoid delay or disruption to our progress, we need your directions by [*insert date*].

Yours faithfully

2.02.3   Letter if contractor says sub-contractor should have found discrepancy

This letter is only suitable for use with DOM/1 or DOM/2

To the Contractor

Dear Sirs

[*Heading*]

Thank you for your letter of the [*insert date*] in which you allege that we should have found the discrepancy in/between [*delete as appriate and insert precise details of the discrepancy*].

Neither clause 4.1.5, nor any other clause in the conditions of sub-contract impose any duty on us to find discrepancies, merely to report discrepancies if we do find them. This is now a principle which we are advised is firmly established by the courts.

We, therefore, reject your allegation and give notice that clause 4.1.5 requires you to issue directions in regard to the discrepancy. This is now a matter of urgency and we are suffering delay and incurring loss and expense which we shall expect to be reimbursed under the provisions of clause 13.

Yours faithfully

2.02.4   Letter if contractor fails to give direction regarding discrepancy
This letter is only suitable for use with DOM/1 or DOM/2
Registered post/recorded delivery

To the Contractor

Dear Sirs

[*Heading*]

We refer to our letter of the [*insert date*] in which we brought a discrepancy to your attention and requested your directions by [*insert date*]. You have not issued any directions and we are now suffering delay and incurring loss and expense. We formally request an extension of time under the provisions of clause 11 of the conditions of sub-contract. In our opinion your failure comes within clause 11.3.1. Treat this also as an application under the provisions of clause 13.1. Further details will be provided in due course.

Yours faithfully

2.02.5   Letter if inconsistency found
This letter is only suitable for use with NAM/SC

To the Contractor

Dear Sirs

[Heading]

We have noted an inconsistency in/between *[delete as appropriate
and specify the inconsistency, giving precise details of drawing
numbers, document references, architect's instructions or contractor's
directions]*

Please now issue your directions in accordance with the provisions
of clause 2.4 of the conditions of sub-contract. If delay and
disruption to the progress of our work are to be avoided, we need
your directions by *[insert date]*.

Yours faithfully

2.02.6   Letter if discrepancy found
This letter is only suitable for use with NSC/C

To the Contractor
(Copy to Architect)

Dear Sirs

[Heading]

In accordance with clause 1.8 of the conditions of sub-contract, we give notice that we have discovered a discrepancy in/between *[delete as appropriate and specify the discrepancy, giving precise details of drawing numbers, numbered document references and architect's instructions]*

Please send this notice to the architect forthwith, requesting instructions under clause 2.3 of the main contract.

In order to avoid delay or disruption to our progress, we need instructions by *[insert date]*.

Yours faithfully

2.02.7   Letter if errors in bills of quantities

This letter is only suitable for use with NAM/SC or NSC/C

To the Contractor
(Copy to Architect and Quantity Surveyor)

Dear Sirs

[*Heading*]

We note the following departure from the specified method of preparation of/error in/misdescription in the bills of quantities: [*delete as appropriate and specify precise details*].

[*When using NAM/SC, add:*]

In accordance with clause 3.4, please issue your directions confirming the architect's instructions in regard to the correction of this problem.

[*When using NSC/C, add:*]

In accordance with clause 1.12.2, it appears that no architect's instruction is necessary. The problem is to be corrected and treated as if it were a variation required by an architect's instruction issued under clause 13.2 of the main contract. We are entitled to have the variation valued by the quantity surveyor under the provisions of clause 4.4 [*substitute clause '4.10' when NSC/A article 3.2 applies*] of the conditions of sub-contract and we should be pleased to receive your confirmation.

Yours faithfully

2.03 Letters of intent

Quite simply, a letter of intent is a letter issued when the other party is not ready to enter into a contract, but he wishes to establish some kind of relationship, usually in order to get something done. Commonly, advance ordering of materials is required which, if left until the sub-contract was executed, would be too late to achieve the kind of progress for which the employer or the contractor is looking. It is called a letter of intent, because the other party usually expresses a firm intention to be contractually bound in the future. Such letters must be treated with great caution and it is advisable to seek proper advice in respect of each such letter received. Some such letters actually create a binding contract, despite the use of the word 'intent'. If there is any doubt, clarification should be sought (**document 2.03.1**)

A nominated sub-contractor under the provisions of JCT 80 will be required to enter into a form of warranty agreement with the employer on form NSC/W, and a named sub-contractor under the provisions of IFC 84 will usually be required to enter into warranty agreement ESA/1 with the employer. Both NSC/W and ESA/1 have provisions for the employer to have materials ordered, work executed or design work carried out before the sub-contract agreement is executed. Therefore, it is unlikely that the employer, or the architect on his behalf, will send a letter of intent in such circumstances.

Under DOM/1 and DOM/2, however, and indeed in respect of other non standard forms, it is common for a sub-contractor to receive a letter of intent in order to start the work. It is common for the sub-contractor to completely finish his work before executing a proper contract. Under such circumstances, it is a matter of looking at all the facts before it is possible to say whether a concluded contract exists. The absence of a concluded contract during a project can have severe consequences for both parties if they fall into dispute.

2.03.1   Letter if doubtful letter of intent received

To the Contractor/the Architect [as appropriate]

Dear Sirs

[Heading]

Thank you for your letter of the *[insert date]* requesting us to *[insert details]*.

We are unsure whether your letter is intended to create a binding contract between us by acceptance of our quotation dated *[insert date]* in the sum of *[insert amount in words and figures]* or whether it is sent as a letter of intent asking us to carry out work for which we shall have to seek reimbursement on a quantum meruit basis.

We shall refer this, as all other documents of this nature, to our legal advisors, but we should be grateful to receive your clarification first, so that time is not wasted.

Yours faithfully

2.04 Quotations and estimates

Although NAM/SC and NSC/C envisage that the sub-contractor will go through a particular tendering procedure, in practice all prospective sub-contractors are apt to find themselves invited to submit a quotation/estimate in a more informal manner. Many sub-contractors firmly believe that to label a price 'estimate' entitles them to adjust the amount demanded after the work is completed. Of course, where a standard form of sub-contract is used, the adjustment of the sub-contract sum must be carried out strictly in accordance with the form of sub-contract being used. Problems arise if there is no form of contract but simply an acceptance by the contractor of the sub-contractor's price. Whether the sub-contractor is entitled to vary the final cost of the work will depend on the precise wording of the sub-contractor's offer and the contractor's acceptance.

In general terms, it can be said that the use of the word 'estimate' does nothing to make the price vague and variable if the price is clearly a figure which is capable of acceptance. It will then become a firm price exactly as if it had been called a 'quotation'. If the sub-contractor is asked to give a rough estimate or budget price which, at the time of the request, is not intended to be firm by either party, he should make the position clear, taking the opportunity to enclose his set of standard conditions, to which the offer should refer (**document 2.04.1**).

2.04.1   Letter giving approximate estimate

To the Contractor

Dear Sirs

[*Heading*]

Thank you for your letter of the [*insert date*] with which you enclosed one copy of [*list the documents enclosed with the letter, giving dates or reference numbers where possible in order to identify them*]. We understand that you only require/It is only possible to give [*delete as appropriate*] an approximate estimate of price or budget figure at this stage.

We enclose a set of our terms and conditions of contract which applies to our offer if you should decide to accept it on this basis.

Yours faithfully

2.05 Signing the sub-contract

Although the sub-contract documents in connection with NAM/SC and NSC/C may be completed by a mixture of people, the party responsible for making sure that they all sit together comfortably when they are ready to sign is the contractor, no matter which sub-contract form is used. Nevertheless, you must carefully check the documents before signing and take a photocopy of the signed documents before returning them to the contractor. Sometimes, you may find that the documents submitted to you have minor (or even major) differences from the documents on which you tendered. It is worth taking some time to check, because once you have signed the sub-contract it will act retrospectively. Clearly, the documents should be exactly the same as the ones on which you tendered otherwise you should waste no time in writing to the contractor (**document 2.05.1**). This is very important if the sub-contract commencement date has already been notified and you are being pressed to start. Do not start on site under any circumstances unless you are sure, with advice if appropriate, that you are already bound into a contract with the contractor on the terms on which you tendered (**document 2.05.2**). You will be so bound if you submitted a form of tender referring to those terms and the contractor sent you an unequivocal acceptance (i.e. one that did not insert any new terms).

Occasionally, you will not be aware of the terms of the sub-contract until, unexpectedly, you are sent a copy and asked to sign. This is common where very informal telephone enquiries and quotations have been used. The golden rule is that if the sub-contract form submitted does not represent the terms you expected, refuse to sign (**document 2.05.3**). It may well be that, by the time you receive the form, a concluded contract, based upon your offer, the contractor's response and your start on site may be in existence. Another common occurrence is that you are unexpectedly asked to complete the document as a deed. This used to involve a seal, but since the Companies Act 1989 and the Law of Property (Miscellaneous Provisions) Act 1989, a document can be executed as a deed provided the correct procedure is observed. This involves the signatures of certain stipulated representatives of the company or, in the case of individuals, the individual and witnesses. **Document 2.05.4** is a set of alternative attestation clauses from NSC/A. It is worth remembering that a contract entered into as a deed has a 12 year limitation period compared to the six year period for contracts executed under hand. The difference may seem trivial when you sign, but it will not be trivial if the contractor writes to you about a serious defect in your work eight years after the issue of the final certificate.

2.05.1   Letter if mistakes in sub-contract documents

To the Contractor

Dear Sirs

[*Heading*]

Thank you for your letter of the [*insert date*] with which you enclosed the sub-contract documents for us to sign/complete as a deed [*delete as appropriate*].

We refer you to [*describe the nature of the error, mistake or inconsistency and page number of document*]. This is not consistent with the documents on which we tendered and we are not prepared to execute them in their present form. We return them herewith and look forward to receiving the corrected documents as soon as possible.

You are pressing us to commence on site and we are advised that we should not start work until a proper binding contract is in place.

Yours faithfully

2.05.2   Letter if mistakes in sub-contract documents and previous acceptance of tender

---

To the Contractor

Dear Sirs

[*Heading*]

Thank you for your letter of the [*insert date*] with which you enclosed the sub-contract documents for us to sign/complete as a deed [*delete as appropriate*].

We are advised that our tender of the [*insert date*] together with your acceptance of the [*insert date*] constitutes a legally binding contract. We are happy to proceed with the work on that basis, but we are not prepared to execute the sub-contract documents, because in the following respects, they do not properly reflect the agreement between us: [*specify*].

We, therefore, return them herewith and we look forward to receiving corrected documents in due course.

Yours faithfully

2.05.3   Letter if asked to sign new terms

To the Contractor

Dear Sirs

[*Heading*]

We are in receipt of your communication dated [*insert date*] which purports to be the terms of the sub-contract between us. These terms are unacceptable, because they are not terms on which we submitted our tender.

We note that you are anxious for us to commence work on site on [*insert date*]. We are prepared to do so only if you first send us a clear and unqualified acceptance of our tender.

In view of our requirement for [*insert number*] days' notice of requirement to commence on site, your acceptance of our tender is now urgent.

Yours faithfully

2.05.4   Alternative attestation clauses from NSC/A

Notes	
	[f1] AS WITNESS THE HANDS OF THE PARTIES HERETO
[f1] For Agreement executed under hand and NOT as a deed.	[f1] Signed by or on behalf of the Contractor_____
	in the presence of:
	[f1] Signed by or on behalf of the Sub-Contractor_____
	in the presence of:

— — — — — — —   Complete under hand (above) or as a deed (below): see NSC/T Part 1, item 12. — — — —

[f2] For Agreement executed as a deed under the law of England and Wales by a company or other body corporate: insert the name of the party mentioned and identified on page 2 and then use *either* [f3] and [f4] *or* [f5].
If the party is an *individual* see note [f6].

[f2]   **EXECUTED AS A DEED BY THE CONTRACTOR** SMITH BROS L^{td}
        hereinbefore mentioned namely_____

[f3]   by affixing hereto its common seal

[f3] For use if the party is using its common seal, which should be affixed under the party's name.

[f4]   in the presence of:

[f4] For use of the party's officers authorised to affix its common seal.

[f5] For use if the party is a company registered under the Companies Acts which is not using a common seal: insert the names of the two officers by whom the company is acting *who MUST be either a director and the company secretary or two directors*, and insert their signatures with 'Director' or 'Secretary' as appropriate.

*   OR ——————————————

[f5]   acting by a director and its secretary* / two directors* whose signatures are here
        subscribed:
        namely_____ A. SMITH

        [Signature] *a. Smith* _____ DIRECTOR

        and_____ B. SMITH

        [Signature] *B. Smith* _____ SECRETARY* /DIRECTOR*

[f2]   **AND AS A DEED BY THE SUB-CONTRACTOR** BLACK BROS L^{td}
        hereinbefore mentioned namely_____

[f3]   by affixing hereto its common seal

[f4]   in the presence of:

[f6] If executed as a deed by an *individual:* insert the name at [f2], delete the words at [f3], substitute 'whose signature is here subscribed' and insert the individual's signature. The individual MUST sign in the presence of a witness who attests the signature. Insert at [f4] the signature and name of the witness. Sealing by an individual is not required.

Other attestation clauses are required under the law of Scotland.

*   OR ——————————————

[f5]   acting by a director and its secretary* / two directors* whose signatures are here
        subscribed:
        namely_____ Y. BLACK

        [Signature] *Y. Black* _____ DIRECTOR

        and _____ Z BLACK

        [Signature] _____ *Z Black* SECRETARY* /DIRECTOR*

* *Delete as appropriate*

2.06 Bonds and warranties

Bonds are always entered into as deeds. They are more usually required from the main contractor than from sub-contractors, but you may be asked to provide a performance bond as a kind of guarantee of proper performance of the sub-contract. If a bond is required, you will have been notified as part of the tender documents, together with an example of the wording required. If you were not so informed, but a bond is still required, you would be prudent to refuse (**document 2.06.1**).

Banks and insurance companies will supply bonds under the proper conditions. In practice, the organization you approach to supply the bond will have its own very clear ideas on what the wording should be. **Document 2.06.2** is an example of a fairly commonly worded and straightforward bond. It is generally said that 'default' bonds are acceptable – because the default must be demonstrated before payment is made – and 'on demand' bonds are not acceptable – because payment is on demand. However, the reality is that the effect of every bond depends upon its own particular wording and there are many subtleties.

Warranties are becoming common. It should be clearly understood that their purpose is no more or less than to give you obligations to parties when, without the warranties, you would not have those obligations to those parties. It is, therefore, better if you do not enter into a warranty agreement with anyone. That said, the standard form warranties ESA/1 and NSC/W confine themselves to clear objectives and even offer some reciprocation to you. You will have very little choice about executing them, but you must always check with your insurers first. You must beware of the non-standard warranties with which you will be bombarded.

There are a number of possibilities if non-standard warranties are concerned. You may have been notified in the tender documentation that you would be expected to enter into a warranty and the text of the warranty may be appended. Alternatively, you may have been notified in the tender documents, but the text may not have been specified. In that situation you are entitled to sign only the text which you approve (**document 2.06.3**). You may never get to that position! Another possibility is that execution of a warranty agreement is not part of the tender requirements and you enter into a contract in which there is no reference to a warranty. Nevertheless, you may receive a request to enter into a warranty. In such circumstances, you should refuse (**document 2.06.4**). It is not always diplomatic to do so and you may offer an alternative and very 'safe' warranty of your own. There is also no reason why you should not agree to enter into a warranty on condition that the contractor, and possibly the employer and architect, also enter into warranties in your favour. There is absolutely no reason why you should not require this as a *quid pro quo*. It is doubtful that they will agree, but their refusal strengthens your own position (**document 2.06.5**).

2.06.1 Letter if bond requested after sub-contract executed

To the Contractor

Dear Sirs

[*Heading*]

We are in receipt of your letter dated [*insert date*] requesting us to supply a bond in the sum of [*insert amount in words and figures*] in accordance with the example attached.

[*If no example in the enquiry document, add:*]

Although we have a contractual obligation to enter into a bond, we have not undertaken to do so in any particular form. We enclose the form of bond which we are prepared to execute and we should be pleased to receive your agreement.

[*Or if no reference to a bond in the enquiry document, add:*]

We have no contractual obligation to enter into a bond and we are advised that we are not now obliged to do so.

Yours faithfully

2.06.2   Example of a typical bond

      BY THIS BOND We _____ whose registered office is
at _____ (hereinafter called 'the Sub-Contractor') and
_____ whose office in the United Kingdom is
at _____ (hereinafter called 'the
Surety') are held and firmly bound unto _____
(hereinafter called 'the Contractor') in the sum of _____
_____ (£   ) for payment of
which sum the Sub-Contractor and Surety bind themselves their successors
and assigns jointly and severally by these presents.
      EXECUTED AS A DEED and dated this _____ day of _____
      WHEREAS by an Agreement dated _____ between the
Contractor and the Sub-Contractor the Sub-contractor contracted with the
Contractor to carry out certain Works as therein mentioned and in conformity
with the provisions of the Contract.
      NOW THE CONDITION of this Bond is such that if the Sub-Contractor shall
duly perform and observe all the terms, provisions, conditions and
stipulations of the Contract on the Sub-Contractor's part to be performed and
observed according to the true purpose, intent and meaning thereof or if in
default by the Sub-Contractor the Surety shall satisfy and discharge the
damages sustained by the Contractor thereby up to the amount of this Bond
then the obligation shall be null and void but otherwise shall be and remain in
full force and effect until _____ but no alteration in terms
of the said Contract made by agreement between the Contractor and the Sub-
Contractor or in the extent of the Works to be carried out thereunder and no
allowance of time by the Contractor under the Contract nor any forebearance
or forgiveness in or in respect of any matter or thing concerning the Contract
on the part of the Contract shall in any way release the Surety from any liability
under this Bond.

The Common Seal of

_____

was hereunto affixed in the presence of:

_____

The Common Seal of

_____

was hereunto affixed in the presence of:

_____

2.06.3 Letter if asked to sign a warranty whose terms not previously agreed

To the Contractor

Dear Sirs

[*Heading*]

We are in receipt of your letter dated [*insert date*] asking us to sign the enclosed warranty in favour of [*insert details*].

Although we agree that by the terms of our sub-contract we undertook to provide a warranty, we did not agree to do so on any particular terms and the terms you specify are not acceptable to us.

May we suggest a meeting to let you know, and to discuss, our detailed objections. Alternatively, we could let you have an example of the form of warranty we would be prepared to sign.

Yours faithfully

2.06.4   Letter if asked to sign a warranty not part of the contract

To the Contractor

Dear Sirs

[*Heading*]

We are in receipt of your letter dated [*insert date*] with which you enclosed a form of warranty in favour of [*insert details*] which you ask us to sign.

We have no obligation under our sub-contract or at law to provide a warranty and, on advice, we respectfully decline to do so.

[*If appropriate, add:*]

If this will cause you particular difficulty, we may be prepared to offer a warranty specially drafted for us on payment of an appropriate sum as consideration.

We look forward to hearing from you.

Yours faithfully

2.06.5   Letter if sub-contractor requires warranties as condition

To the Contractor

Dear Sirs

[*Heading*]

We are in receipt of your letter dated [*insert date*] with which you enclosed a warranty in favour of [*insert details*] which you ask us to sign.

Our sub-contract places us under no obligation to sign such a warranty, but we may be prepared to do so if you/the architect/the employer [*delete as appropriate*] will sign the enclosed specially drafted warranty in our favour. You will note that it does no more than protect our reasonable expectations under the sub-contract.

Yours faithfully

Chapter 3

Named Persons as Sub-Contractors

3.01 Introduction

The 'naming' provisions in the IFC 84 form of main contract are a kind of watered down version of the JCT 80 nomination provisions. They are contained in clause 3.3. You may be named in one of two ways. In the first instance, the work may be included in the main contract documents and is to be priced by the main contractor and carried out by a named person. In this case, you are named in the contract documents and when the contractor tenders for the project, you are part and parcel of the project. If the contractor has had dealings with you before and he would rather not deal with you again, his only option is to refrain from tendering. If he does tender, you are part of the package. In the second instance, the architect may name you in an architect's instruction which he has issued regarding the expenditure of a provisional sum. In this case, the contractor is entitled to object (see Section 3.03).

IFC 84 deals with the situation if the sub-contract cannot be executed (clause 3.3.1), if the contractor objects (clause 3.3.2), if the named person's employment is determined (clause 3.3.3), the aftermath of determination (clause 3.3.6), and design liability (clause 3.3.7).

3.02 Documentation NAM/T and NAM/SC

Whether the architect intends to name you in the main contract documents or in an instruction in connection with the expenditure of a provisional sum, he must use the JCT Form of Tender and Agreement NAM/T. The architect must complete section I, the invitation to tender. An example is shown in **document 3.02.1**. Its purpose is to give you basic information so that you can tender on a proper basis. It is not unknown for the architect to accidently omit some portions and you must be vigilant. You must not attempt to fill in the blanks yourself. Rather, you must write to the architect, returning NAM/T and requesting its proper completion (**document 3.02.2**).

Once you are satisfied that section I has been properly completed, you must complete section II, the tender. An example is shown in **document 3.02.3**. You must read the notes with great care, because some items are in the alternative. Different offers are to be completed depending upon whether you are to be named in the main contract documents or in an instruction. Despite the statement that your offer will remain open for acceptance for a stipulated period, you may withdraw your offer at any time before acceptance (**document 3.02.4**). The architect may not appreciate this action and mistakenly consider that you are in breach of contract. You must clarify the situation (**document 3.02.5**). You have the opportunity in this section to qualify your offer and to stipulate programme requirements.

It is important to remember that although the offer is addressed to the employer and to the contractor, it is only the contractor who can accept it. You may well receive a letter from the architect to the effect that your offer is accepted, but it is of no effect and you should waste no time in informing him so that the tender can be accepted properly (**document 3.02.6**). The contractor may simply write to you informing you that your tender is satisfactory and sending you his 'official order'. Reject it in favour of the Articles of Agreement which form section III of NAM/T. 'Official orders' often have unpleasant conditions attached (**document 3.02.7**).

On completion of section III (see example in **document 3.02.8**), you will be in contract with the main contractor in a form which has a clause (Article 1.2) incorporating NAM/SC conditions. It is not usual for sub-contracts to physically incorporate NAM/SC, therefore it is a good idea to have an office copy of the conditions for reference purposes. Many sub-contractors fail to secure their rights simply because they do not know the conditions under which they have contracted. Of course, the prudent contractor will have completely familiarized himself with the conditions before tendering.

3.02.1 NAM/T Section I

JCT

Incorporating Amendments 1: 1986, 2: 1987, 3: 1988, 4: 1990 and 5: 1991   Tender and Agreement NAM/T

Form of Tender and Agreement

for a person to be named by the Employer as a sub-contractor under the JCT Intermediate Form of Building Contract 1984 Edition (IFC 84), clause 3·3

*Should there be a separate agreement between the Employer and the Sub-Contractor relating to such matters as are referred to in clause 3·3·7 of the main contract conditions (design etc.), it should **not** be attached to the Form of Tender and Agreement, either where the Form is included in the main contract documents or where it is included in an instruction of the Architect/the Contract Administrator for the expenditure of a provisional sum.*

Notes on completion

[a] The Architect/The Contract Administrator to complete the whole of Section I.

[b] Insert name and address of sub-contractor.

[c] Insert the name and address of the person to whom NAM/T should be returned by the sub-contractor after he has completed Section II: this should normally be the Architect/the Contract Administrator.

[d] The Architect/The Contract Administrator should sign the Invitation to Tender and, unless already given at [c], insert his name and address.

[a] ## Section I – Invitation to tender

[b] To   Acme TV Aerials Ltd

    Nefhar Works

    Little Industrial Estate

    Little Nefhar

    Blankshire

You are invited to tender, as a person who is to be employed by the Contractor under a sub-contract in accordance with the Form of Agreement in Section III hereof, a VAT-exclusive Sub-Contract Sum for executing the Sub-Contract Works referred to below by completing [c] SECTION II and returning the whole form to

    Inigo Waffle & Associates

    Architects

Your tender must not vary any of the matters set out in SECTION I of this Form.

The documents listed below, hereinafter called 'the Numbered Documents', are enclosed herewith:

    Drawing Numbers: ABC/3A, 4B, 7, 8D and 9A.

    Specification SPC/AER/1

Should it be decided to name you as the person to execute the Sub-Contract Works, you will be required to provide the following, hereinafter called 'the Priced Documents':

* a priced copy of the Specification
~~a priced copy of the Schedule of Works~~
~~a priced copy of the Bills of Quantities~~ ×
~~a Sub-Contract Sum Analysis~~
~~a Schedule of Rates~~

and any such priced copy of a Numbered Document shall be deemed to replace the unpriced copy.

[d] Signed _____

Name   Inigo Waffle & Associates

Address   Drawtooher Close

       Little Nefhar

Date   21 March   19 94

3.02.1   NAM/T Section I *Continued*

Notes on completion

Section I – Invitation to tender

[e] Insert the same description as in IFC 84, 1st Recital.

[e] Main Contract Works and location_____

Housing Development for Zippo Inc.

Withering Heights

Off Dark Lane

Little Nefhar

Blankshire

_____Job reference__ABC__

Particulars of the Sub-Contract Works__Supply and erection of__

patent TV aerial system to each of 52 houses.

Names and address of:

Employer:	Tel No:
Zippo Inc	0001 1010101
Zippo House	
Zippo Street	
Nefhar	

[f] Main Contract clause 8-4.

[f] *The Architect/The Contract Administrator: Tel No:

  Inigo Waffle & Associates   0001 20202
  Drawtooher Close
  Little Nefhar

Quantity Surveyor: Tel No:
  William raites   0001 30303
  Dotton Place
  Timesing

[g] Where item 12(a) applies the Contractor will not have been appointed: instead the names of contractors who will be invited to tender should be set out.

[g] Main Contractor: Tel No:

  Gerrybuilders Ltd   0001 40404
  Downther Road
  Laffin

3.02.1   NAM/T Section I *Continued*

Notes on completion

Section I – Invitation to tender

MAIN CONTRACT INFORMATION

1	Form of Main Contract	JCT Intermediate Form of Building Contract 1984 Edition (IFC 84) incorporating Amendments 1 to 6

2   Main Contract alternative or optional provisions (See also item 6)

*Specification/~~Schedules of Work/Bills of Quantities~~: 1st Recital

*alternative A/~~alternative B~~: 2nd Recital

*The Architect/~~the Contract Administrator~~: Article 3

3   Any changes from printed Form of Main Contract Conditions identified at item 1:

[h] See also item 13 page 6.

4   [h] Execution of Main Contract: *is/~~is not~~* under hand/~~as a deed~~

5   Inspection of Main Contract: the unpriced *Specification/~~Schedules of Work/Bills of Quantities~~ and the Contract Drawings may be inspected at:
(where item 12(a) applies the documentation for the Main Contract may not be yet be available)

The Architect's office by appointment

6   Main Contract: Appendix and entries therein as amended by amendments 1 to 6
(where item 12(a) applies it is intended that the following entries will be made in the Appendix to the Main Contract. When item 12(b) applies the following are the entries to the Appendix to the Main Contract.)

IFC84 Clause

2·1	(i) Date of Possession	7 February 1994
2·1	(i) Date for Completion	31 December 1995
2·4·10 and 2·4·11	Extension of time for inability to secure essential labour or goods or materials	* Clause 2·4·10 *(Labour)* applies/~~does not apply~~ * Clause 2·4·11 *(Goods or materials)* applies/ ~~does not apply~~
2·2 and 2·4·14 and 4·11(a)	Deferment of the Date of Possession	* Clause 2·2 applies/~~does not apply~~ Where clause 2·2 applies, _____6_____ weeks (period not to exceed 6 weeks)
2·7	Liquidated damages	at the rate of £ 520.00 per week

(i) See item 10 page 5.

*delete as applicable

3.02.1   NAM/T Section I *Continued*

Section I – Invitation to tender

MAIN CONTRACT INFORMATION continued

6	Main Contract: Appendix and entries therein as amended by Amendments 1 to 6

IFC84 Clause

2·10	Defects liability period (if none stated is 6 months from the day named in the certificate of Practical Completion of the Works)	12 months
4·2	Period of interim payments if interval is not one month	one month
4·9(a) and C7	Supplemental Condition C: Tax etc. fluctuations	Percentage addition _____ 10 _____ %
4·9(b)	Formulae fluctuations (not applicable unless Bills of Quantities are a Contract Document)	Supplemental Condition D ~~applies~~/does not apply
D1	Formula Rules (only where Supplemental Condition D applies)	rule 3: Base Month _____19____ [A] rule 3: Non-Adjustable Element _____(not to exceed 10%) rules 10 and 30(i) * Part I/Part II of Section 2 of the Formula Rules is to apply
5·5	Value Added Tax: Supplemental Condition A	Clause A1·1 of Supplemental Condition A ~~applies~~/does not apply
6·2·1	Insurance cover for any one occurrence or series of occurences arising out of one event	£ 2,000,000.00
6·2·4	Insurance – liability of Employer	Insurance may be required ~~is not required~~ Amount of indemnity for any one occurrence or series of occurrences arising out of one event £ 200,000.00
6·3·1	Insurance of the Works – alternative clauses	* Clause 6·3A/~~6·3B/6·3C~~ applies
6·3A·1* 6·3B·1* 6·3C·2*	Percentage to cover professional fees	12 _____ %
6·3A·3·1	Annual renewal date of insurance as supplied by Contractor	1 April
6·3D	Insurance for Employer's loss of liquidated damages – clause 2·4·3	* Insurance ~~may be required~~/is not required
6·3D·2	Period of time	NA
8·3	Base Date	1 January 1994
9·1	Appointor of Arbitrator	President or a Vice-President: * Royal Institute of British Architects ~~Royal Institution of Chartered Surveyors~~ x ~~Chartered Institute of Arbitrators~~ x
9·6	Reference to Arbitrator under NAM/SC	* clause 9·6 applies/~~does not apply~~

[A] Only applicable where the Employer is a Local Authority

3.02.1   NAM/T Section I *Continued*

Notes on completion

Section I – Invitation to tender

MAIN CONTRACT INFORMATION continued

[j] This information, unless included in the Numbered Documents (see page 1), should be given, eg. by repeating it here or by attaching a copy of the relevant part of the Main Contract Documents.

7   Obligations or restrictions which are or will be imposed by the Employer not covered by the Main Contract Conditions (eg. those which are or will be included in the Specification/the
[j] Schedules of Work/the Contract Bills, or are in Variation instructions):

   See sheet A attached

8   Order of Works: Employer's requirements affecting the order of the Main Contract Works (if any):

   See drawing number 3A attached

9   Location and type of access:

   Off Dark Lane - any vehicle

Where item 12(b) or 12(c) applies:

10   New dates for possession or completion where these have been altered from the original dates stated in the Main Contract Appendix reproduced at item 6 pages 3 and 4:

   As Appendix

11   Other relevant information (if any) relating to the Main Contract:

3.02.1 NAM/T Section I *Continued*

Notes on completion

Section I – Invitation to tender

SUB-CONTRACT INFORMATION

12   NAM/T with Section I and II completed, the Priced Documents and such of the Numbered Documents as are not Priced Documents will, if approved, be:

    *(a) ~~included in the Main Contract Documents for pricing by a contractor~~ ~~(see IFC 84, 1st Recital)~~

  or *(b)   included in an instruction of the Architect/the Contract Administrator to the Contractor as to the expenditure of a provisional sum (see IFC 84, clause 3·3·2);

  or *(c) ~~included in an instruction of the Architect/the Contract Administrator as to the expenditure~~ ~~naming a replacement sub-contractor (see IFC 84, clause 3·3·3(d))~~

13   The Named Sub-Contractor will be required to enter into a sub-contract in accordance with the Form of Agreement in Section III with the Contractor selected by the Employer to execute the Main Contract Works. (Where item 12(b) or (c) applies the Contractor will first have the right to make a reasonable objection to entering into a sub-contract with the Named Sub-Contractor.) Unless otherwise stated below the Form of Agreement shall be entered into in the same manner as the main contract (see item 4 page 3).

[k] See items 6 and 10.

14   [k]   The Main Contract Appendix and entries therein will, where relevant, apply to the Sub-Contract unless otherwise specifically stated here:

    _ _ _ _ _ _ _ _ _ _ _ _ _ _

[l] The actual date or dates for commencement of the Sub-Contract Works should be settled by the Contractor and Sub-Contractor taking into account any period of notice required to be given to the Sub-Contractor to commence work on site to be set out in Section II, item 1(3). (See Section II, item 1).

15   [l]   The dates between which it is expected that the Sub-Contract Works can be commenced on site:

to be between   1 September 1994

and   1 November 1994

Period required by the Architect/the Contract Administrator to approve drawings after submission   5 days

*delete as applicable

3.02.1   NAM/T Section I *Continued*

Notes on completion

Section I – Invitation to tender

SUB-CONTRACT INFORMATION continued

[m] The Architect/The Contract Administrator should complete the appropriate part of item 16 and delete the other part.

16 [m]   **Sub-Contract Fluctuations**

1   *NAM/SC – clause 33 – Contributions, levy and tax fluctuations will apply

Duties and taxes on fuels: * included/~~excluded~~ (NAM/SC 33·2·1)

Base Date (NAM/SC 33·6·1) _____ 4 April 1994

Percentage addition to fluctuation payments or allowances (NAM/SC 33·7)
10 _____ %

2   *NAM/SC – clause 34 – Formula adjustment will apply

Sub-Contract/Works Contract Formula Rules are those dated (NAM/SC 34·1)

_____ 19___

*Part I/Part III of these Rules applies

[n] Only applicable where the Employer is a Local Authority.

[n] Non-Adjustable Element _____ % (not to exceed 10%) (NAM/SC 34·3·3)

NAM/SC Formula Rules

Definition of Balance of Adjustable Work – any measured work not allocated to a Work Category (rule 3)

_____

_____

_____

_____

[o] The Base Month should normally be the calendar month prior to that which this Tender is due to be returned.

[o] Base Month (rule 3) _____

Base Date _____

[p] If not completed by the Architect/the Contract Administrator to be completed by the Sub-Contractor in Section II, item 4.

[p] Method of dealing with 'Fix-only' work (rule 8)

[p] Part I of Formula Rules only: the Work Categories applicable to the Sub-Contract Works (rule 11)

[p] Part III of Formula Rules only: Weightings of labour and materials – Electrical Installations or Heating, Ventilating and Air Conditioning Installations (rule 43)

	Labour	Materials
Electrical	_____ %	_____ %

[q] If both specialist engineering formulae apply to the Sub-Contract the percentages for use with each formula should be inserted and clearly identified. The weightings for sprinkler installations may be inserted where different weightings are required.

[q] Heating, Ventilating and Air Conditioning _____ % _____ %

[q] _____ _____ % _____ %

[p] Adjustment shall be affected (rule 61a)

*upon completion of manufacture of all fabricated components

~~*upon delivery to site of all fabricated components~~

*delete as applicable

3.02.1 NAM/T Section I *Continued*

Section I – Invitation to tender

SUB-CONTRACT INFORMATION continued

[r] Any other additional attendance and any other special requirements or any variation to those set out by the Architect/the Contract Administrator here should **be detailed by the Sub-Contractor in Section II, item 3.**

17   The attendance to be provided by the Contractor free of charge to the Sub-Contractor will be as stated in NAM/SC clause 25·1. The following additional attendance and/or special
[r]   requirements will also be provided free of charge to the Sub-Contractor (NAM/SC 25·3):

NONE

18   Settlement of disputes – arbitration – appointer (if no appointor is selected the appointer shall be the President or a Vice-President, Royal Institution of Chartered Surveyors)

35·1   President or a Vice-President:
~~*Royal Institution of Chartered Surveyor~~
*Royal Institute of British Architects
~~*Chartered Institute of Arbitrators~~

3.02.2 Letter if architect omits portions of NAM/T
This letter is only suitable for use with NAM/SC

To the Architect

Dear Sirs

[Heading]

Thank you for your letter of the *[insert date]* inviting us to tender for *[insert details]*. We shall be happy to submit a tender.

On looking through the form NAM/T, we noticed that there were a number of instances in section 1 (invitation to tender) where spaces have been left blank. We have taken a copy of the form and we are preparing our tender on the basis of the information included, but we return the original form herewith. Perhaps you will fill in the places we have flagged and return it to us as soon as possible so that we can complete our tender and get it to you within the timescale you require.

We are anxious to be of service, so do not hesitate to telephone if we can assist further.

Yours faithfully

3.02.3 NAM/T Section II

Notes on completion

[a] Sub-Contractor to complete the whole of Section II. See page 12 for counter-signature by the Architect/the Contract Administrator.

[s] Section II – Tender by sub-contractor

To: the Employer and Contractor

In response to the invitation in Section I

We ___Acme TV Aerials Ltd___

of ___Nefhar Works, Little Ind. Estate, Little Nefhar___

___Blankshire___ Tel No: ___0001 50505___

have duly noted the information therein contained and now OFFER to carry out and complete, as a named person to be employed by the Contractor as a sub-contractor, the Sub-Contract Works identified in the Numbered Documents listed in Section I and in accordance with the entries we have made in items 1-6 of this Section for

the VAT-exclusive sum (hereinafter referred to as 'the Sub-Contract Sum') of £ ___7,800.00___

and

[t] Where item 12(a) of Section I applies, the Sub-Contractor should note the provisions in clause 3·3·1 of the Main Contract Conditions.

[t] where item 12(a) of Section I applies, to conclude a sub-contract with the Contractor by completing Section III hereof within 21 days of the Contractor entering into the Main Contract with the Employer; or

[u] Where item 12(b) or 12(c) of Section I applies the Sub-Contractor should note the provisions in clause 3·3·2 of the Main Contract Conditions. The Section II items set out below, as completed, are to be counter-signed by the Architect/the Contract Administrator on page 12.

[u] where item 12(b) or 12(c) of Section 1 applies, to conclude a sub-contract with the Contractor by completing Section III hereof immediately on the issue of the instruction by the Architect/the Contract Administrator under the Main Contract Conditions clauses 3·3·2 or 3·3·3(a) and 3·8 but subject to the right of the Contractor to make a reasonable objection to entering into such a sub-contract within 14 days of the date of issue of the instruction.

AND with: the **daywork percentages** set out below; and

the **fluctuation provisions** set out at Item 4;

any **additional attendance or special requirements** set out at Item 5.

[v] See note [dd] page 12.

[v] **This Tender, subject to any extension of the period for its acceptance, is withdrawn if not**

accepted by the Contractor within ___12___ **(weeks) of the date of this Tender.**

The daywork percentages are (NAM/SC 16·2·5):

[w] Where more than one Definition will be relevant set out percentage additions applicable to each such Definition. The four Definitions which may be identified are: those agreed between the Royal Institution of Chartered Surveyors and the Building Employers Confederation; the Royal Institution and the Electrical Contractors Association; the Royal Institution and the Electrical Contractors Association of Scotland; and the Royal Institution and the Heating and Ventilating Contractors Association.

[w] Definition	Labour %	Materials %	Plant %
* RICS/BEC			
* RICS/ECA	220	17	17
* RICS/ECA (Scotland)			
* RICS/HVCA			

Daywork percentages take into account the 2½% cash discount allowable to the Contractor under Sub-Contract NAM/SC (NAM/SC 19·3·2 and 19·8·2).

3.02.3   NAM/T Section **II** *Continued*

Notes on completion

Section II – Tender by sub-contractor

ITEMS TO BE COMPLETED BY THE SUB-CONTRACTOR

[x] See also Section I,
item 15.

1   [x]   The periods required (NAM/SC 12·1):

   (1)   for submission of any further sub-contractor's drawings etc. (co-ordination, installation, shop or builder's work or other as appropriate)

   _____2_____weeks from receipt of all necessary information therefor

   (2)   for execution of any Sub-Contract Works off-site

   _____1_____weeks from approval of drawings or establishment of site-dimensions as appropriate

   (3)   for notice to commence work on-site

   _____2_____weeks

   (4)   for execution of Sub-Contract Works on-site

   _____52_____weeks from the date stated in the notice under (3) above.

2   Insurance cover for any one occurrence or series of occurrences arising out of one event (NAM/SC amended by Amendment 1: issued November 1986, clause 8·2)

[y] Must not be less than
the amount inserted in Main
Contract Appendix:
see Section I, item 6.

[y] £___2,000,000.00   (Two million pounds only)___

[z] See item 17 and
sidenote [r] in Section I.

3   [z]   Any additional attendance and/or any other special requirements which vary or add to those referred to in Section I, item 17, and which the Sub-Contractor requires the Contractor to provide free of charge should be set out here in reasonable detail (NAM/SC 25·3):

   – – – – – – – –

3.02.3   NAM/T Section II *Continued*

Notes on completion

Section II – Tender by sub-contractor

ITEMS TO BE COMPLETED BY THE SUB-CONTRACTOR continued

4   **Sub-Contract Fluctuations**

Fluctuations will be in accordance with clauses 33 or 34 of NAM/SC as stated in Section I, item 16.

Where it is stated that clause 33 applies, a list of materials, goods, electricity and fuels (where fuels are to be included: see NAM/SC clause 33·2·1) is attached on a separate sheet.

Where it is stated that clause 34 applies, the following will apply (NAM/SC 34):

Fluctuations – articles manufactured outside the United Kingdom. List of market prices of such articles which the Sub-Contractor is required by the Sub-Contract Documents to purchase and import (see Formula Rules, rule 4 (ii)) – is attached on a separate sheet (NAM/SC 34·4).

[aa] To be completed only to the extent that the Architect/the Contract Administrator has not completed these items in Section I, item 16.

[aa]   Method of dealing with 'Fix-only' work (rule 8)

[aa]   Part I only: the Work Categories applicable to the Sub-Contract Works (rule 17a)

[aa]   Part III only: Weightings of labour and materials – Electrical Installations of Heating, Ventilating and Air Conditioning Installations (rule 43)

	Labour	Materials
Electrical	_____%	_____%

[bb] If both specialist engineering formulae apply to the Sub-Contract the percentages for use with each formula should be clearly identified. The weightings for sprinkler installations may be inserted where different weightings are required.

[bb]   Heating, Ventilating and Air Conditioning _____% _____%

_____ _____% _____%

[aa]   Adjustment shall be effected (rule 61a)

*upon completion of manufacture of all fabricated components

*upon delivery to site of all fabricated components

Part III only: Structural Steelwork Installations (rule 64):

(i)   Average price per tonne of steel delivered to fabricator's work

£ _____

(ii)   Average price per tonne for erection of steelwork

£ _____

Catering Equipment Installations (rule 70a):

apportionment of the value of each item between

cc] Insert values on a separate sheet.

[cc]   (i)   materials and shop fabrication
[cc]   (ii)   supply of factor items
[cc]   (III)   site installations

3.02.3   NAM/T Section II *Continued*

Section II – Tender by sub-contractor

ITEMS TO BE COMPLETED BY THE SUB-CONTRACTOR continued

5   Any other matters (eg. special conditions or agreements on employment of labour, limitations on working hours) to be set out here:

- - - - - - - - - -

Finance (No. 2) Act 1975 – Statutory Tax Deduction Scheme (NAM/SC 18A)

6   The evidence to be produced to the Contractor for the verification of the Sub-Contractor's tax certificate

expiry date_____5 April_____19_95_

will be:

Current tax certificate

[dd] Before counter-signing this Tender and passing it to the Contractor the Architect/ the Contract Administrator should:
(1) obtain if necessary extension of the period for acceptance of the Tender stated on the first page of Section II;
(2) delete (A) or (B) as appropriate;
(3) inform the Sub-Contractor of any changes to the information in Section I, which should be recorded against the relevant item in Section I and initialled by the Sub-Contractor; and
(4) obtain from the Sub-Contractor the Priced Documents identified on page 1.

Signed by or on behalf
of the Sub-Contractor ___ O N Thiles

Dated____25 March_____19_94_

*(A) ~~This is the Tender document referred to in the Articles of Agreement signed by the Contractor~~

or *(B) This is the Tender Document referred to in

Instruction No ____24____, dated____14 April 1994____19_____

[dd] Signed by or on behalf
of the Architect/the
Contract Administrator _____

Dated____14 April_____19_94_

3.02.4 Letter if withdrawing offer
This letter is only suitable for use with NAM/SC
By fax and post

To the Contractor [if known] and the Employer
(copy to the Architect)

Dear Sirs

[*Heading*]

We refer to our tender on NAM/T, section II, dated [*insert date*] in the sum of [*insert amount in words and figures*] for the execution of the above work.

Our tender is hereby withdrawn and it is no longer capable of acceptance.

Yours faithfully

3.02.5   Letter if architect thinks withdrawal is breach of contract

To the Architect/the Contractor/the Employer [as appropriate]

Dear Sirs

[*Heading*]

Thank you for your letter of the [*insert date*]. We note you view that we are not entitled to withdraw our tender until the period specified in NAM/T, page 9, has elapsed.

We assure you that, under English law, we are so entitled unless we have received some consideration for leaving the tender open. That is not the case in this instance. For the same reason, reference to breach of contract is irrelevant, because there is, as yet, no contract to breach.

Having said that, we have no desire to cause you embarrassment; the withdrawal was a result of serious commercial factors. We should be delighted to discuss the matter with you in an attempt to reach a solution satisfactory to all parties.

We look forward to hearing from you.

Yours faithfully

3.02.6 Letter if architect tries to accept tender
This letter is only suitable for use with NAM/SC

To the Architect

Dear Sirs

[*Heading*]

Thank you for your letter of the [*insert date*] by which you
purport to accept our NAM/T, section II, tender dated [*insert date*]
in the sum of [*insert amount in words and figures*].

May we refer you to the wording of NAM/T which makes clear
that it can only be accepted by the contractor entering into the
agreement in section III.

We expect this is an oversight and we look forward to receiving
section III from the contractor so the sub-contract may be properly
executed.

Yours faithfully

3.02.7   Letter if contractor sends 'official order'
This letter is only suitable for use with NAM/SC

To the Contractor

Dear Sirs

[*Heading*]

Thank you for your 'official order' number [*insert number*] of the [*insert date*] by which you purport to accept our NAM/T, section II, tender dated [*insert date*] in the sum of [*insert amount in words and figures*].

We refer you to NAM/T, which indicates that our tender can only be accepted if you enter into a sub-contract with us as set out in section III.

We expect that this is an oversight on your part and we look forward to receiving a completed section III ready for execution.

Yours faithfully

3.02.8 NAM/T Section III

Section III – Articles of Agreement

This Agreement

is made the _____21_____ day of _____April_____ 19_94_

between the Contractor and the Sub-Contractor named or referred to in the foregoing Sections I and II.

Whereas

1st   The Contractor desires to have executed the works referred to in the aforementioned Section I and described in the Numbered Documents identified in that Section;

2nd   The Sub-Contractor has submitted the Tender set out in the aforementioned Section II and the Priced Documents identified in Section I;

3rd   The Numbered Documents (which as stated in Section I include the Priced Documents), the Schedule of Rates and the Contract Sum Analysis, as appropriate, have been signed by the Contractor and the Sub-Contractor and attached hereto;

4th   At the date of this Agreement:

    (A)   the Sub-Contractor is/is not* the user of a current sub-contractor's tax certificate under the provisions of the Finance (No 2) Act 1975 (hereinafter called 'the Act') in one of the forms specified in Regulation 15 of the Income Tax (Sub-Contractors in the Construction Industry) Regulations, 1975, and the Schedule thereto (hereinafter called 'the Regulations'), where the words 'is not' are deleted, clause 18A of the Sub-Contract Conditions referred to in Article 1·2 shall apply to the Sub-Contract and clause 18B of the said Sub-Contract Conditions shall not apply; where the word 'is' is deleted, clause 18B shall apply to the Sub-Contract and clause 18A·2 to ·8 shall not apply;

    (B)   The Contractor is/is not* the user of a current sub-contractor's tax certificate under the Act and the Regulations;

    (C)   The Employer under the Main Contract is/is not* a 'contractor' within the meaning of the Act and the Regulations.

3.02.8   NAM/T Section III *Continued*

Section III – Articles of Agreement

Now it is hereby agreed as follows

Article 1

Sub-Contractor's
Obligations

1·1 For the consideration mentioned in Article 2 the Sub-Contractor shall, upon and subject to the Sub-Contract Documents, namely this Tender and Agreement NAM/T, the Sub-Contract Conditions and the Numbered Documents, carry out and complete the Sub-Contract Works in accordance with the requirements, if any, of the Contractor for regulating the due carrying out of the Works which are agreed by the Sub-Contractor, initialled by the Contractor and Sub-Contractor, attached hereto and incorporated herein, provided that such requirements shall not alter any item set out in Section I or II of NAM/T.

Sub-Contract
Conditions

1·2 The Sub-Contract Conditions are those set out in the 'Sub-Contract Conditions NAM/SC' 1984 Edition (incorporating amendments 1 to 6) issued by the Joint Contracts Tribunal, which shall be deemed to be incorporated herein.

Clause 17A and clause 17B:	Value Added Tax		Clause *17A/17B will apply
		[dd·1]	Clause 17A·5 *applies/does not apply
		[dd·1]	*Clause 17B·5 *applies/does not apply

Article 2

Sub-Contract Sum

The Contractor shall pay to the Sub-Contractor the VAT-exclusive sum of

£ __7,800.00__

__Seven Thousand Eight Hundred Pounds Only__

_____ (words)

(hereinafter referred to as 'the Sub-Contract Sum') or such other sum as shall become payable in accordance with the Sub-Contract.

Article 3

Adjudicator and
Trustee-Stakeholder

3·1 [ee] The name and address of the Adjudicator for the purposes of clause 22·1·2 of NAM/SC is:

I M Partial Esq

Arbitration Road

Greater Nefhar, Blankshire

3·2 [ee] The name and address of the Trustee-Stakeholder for the purposes of clause 22·3·1·2 of NAM/SC is:

The Acme Bank

High Street

Nefhar, Blankshire

Article 4

Settlement of
disputes – arbitration

[ff] Any dispute or difference between the Contractor and the Sub-Contractor shall be referred to arbitration in accordance with and subject to the provisions of the Sub-Contract.

[dd·1] Clause 17A·5 or clause 17B·5 can only apply where the Sub-Contractor is satisfied at the date the Sub-Contract is entered into that his output tax on **all** supplies to the Contractor under the Sub-Contract will be at either a positive or a zero rate of tax. Some supplies by the Contractor to the employer are zero rated by a certificate in statutory form. Only the person holding the certificate, usually the Contractor, may zero rate his supply. Sub-contract supplies for a main contract zero rated by certificate are standard rated: see the VAT leaflet 708 revised 1989.

[ee] For these provisions to be effective the names and addresses identifying the Adjudicator and the Trustee/Stakeholder should be inserted at the time of entering into this Agreement.

[ff] The main provisions dealing with arbitration are to be found in clause 35 of the Sub-Contract Conditions; but references to arbitration may also arise under clauses 18A or 18B, and clause 22.

*delete as applicable

3.02.8 NAM/T Section III *Continued*

Section III – Articles of Agreement   **Tender and Agreement NAM/T**

Notes

[A1] For Agreement executed under hand and NOT as a deed.

[A1] **AS WITNESS THE HANDS OF THE PARTIES HERETO**

[A1] Signed by or on behalf of the Contractor _____ *EJ Peter*

in the presence of: *G. Gerry.*

[A1] Signed by or on behalf of the Sub-Contractor _____ *O N Ihiles*

in the presence of: *I. Drive*

[gg] Complete under hand *(above)* or as a deed *(below)* as applicable: see items 4 and 13 in Section I.

[A2] For Agreement executed as a deed under the law of England and Wales by a company or other body corporate: insert the name of the Contractor/Sub-Contractor and then use *either* [A3] and [A4] *or* [A5]. If the party is an *individual* see note [A6].

[A3] For use if the party is using its common seal, which should be affixed under the party's name.

[A4] For use of the party's officers authorised to affix its common seal.

[A2] **EXECUTED AS A DEED BY THE CONTRACTOR**
hereinbefore mentioned namely _____

* [A3] by affixing hereto its common seal

[A4] in the presence of:

* OR

[A5] For use if the party is a company registered under the Companies Acts which is not using a common seal: insert the names of the two officers by whom the company is acting *who MUST be either a director and the company secretary or two directors*, and insert their signatures with 'Director' or 'Secretary' as appropriate.

[A5] acting by a director and its secretary* / two directors* whose signatures are here subscribed:
namely _____

[Signature] _____   *DIRECTOR*

and _____

[Signature] _____   *SECRETARY* /DIRECTOR**

[A2] **AND AS A DEED BY THE SUB-CONTRACTOR**
hereinbefore mentioned namely _____

* [A3] by affixing hereto its common seal

[A4] in the presence of:

[A6] If executed as a deed by an *individual:* insert the name at [A2], delete the words at [A3], substitute 'whose signature is here subscribed' and insert the individual's signature. The individual MUST sign in the presence of a witness who attests the signature. Insert at [A4] the signature and name of the witness. Sealing by an individual is not required. See specimen overleaf.

Other attestation clauses are required under the law of Scotland.

* OR

[A5] acting by a director and its secretary* / two directors* whose signatures are here subscribed:
namely _____

[Signature] _____   *DIRECTOR*

and _____

[Signature] _____   *SECRETARY* /DIRECTOR**

* *Delete as appropriate*

3.03   Main contractor's right of objection

Clause 3.3.2 of IFC 84 deals with the situation if the architect wishes to issue an instruction naming you to the main contractor. The architect must have gone through the NAM/T procedure to invite your tender, but at the end of section II where the architect has to signify his agreement, he will delete part (A) and insert the name of the instruction and the date of issue. The instruction must also have the numbered documents attached.

On receipt of the instruction, the contractor has 14 days in which to register a reasonable objection to entering into contract with you. It is comparatively rare for this to happen and there is very little that you can do about it. If you are to be the subject of an instruction, you will know the name of the contractor when you tender. If you have any reservations, you need not tender. On the other hand, the contractor's first knowledge will be when he reads the instruction unless the architect has already telephoned him to get his reaction in advance. In the first instance, it is the architect who will decide whether the contractor's objection is reasonable. If he does not agree with the contractor, either party may seek arbitration. That is not a very good idea at this stage in a contract. So if the contractor has some objection to working with you, his view is likely to prevail.

3.04   Substituted sub-contractors

Determination is dealt with in Chapter 10, but one aspect is worth mentioning here. Under clause 3.3.3 of IFC 84, if the employment of a named person is determined for any reason, the architect has the option of naming another sub-contractor to complete the sub-contract work. If you are that sub-contractor, you will be named in an architect's instruction after the proper tendering procedure has been completed. The contractor will have the right to object. In the circumstances which exist after a named person's employment has been determined, it is unlikely that the contractor will object to the substitute sub-contractor. All parties will be anxious to lose as little time as possible.

There are no particular points to watch except that under clause 3.3.4 (a), the contractor is entitled to have his contract sum increased by the additional cost of employing a substitute sub-contractor. This right is restricted to the extent that there are any defects in the first sub-contractor's work which need repair. The contract clearly intends that the contractor must stand that cost himself, because it is his responsibility that the work is properly constructed. The contractor may attempt to pass these costs to you. Remember that you are entitled to be paid in accordance with the terms of your sub-contract. You are not responsible for the financial implications of another's defects unless you have been expressly requested to include for the rectification in your price (**document 3.04.1**). In practice, it is unlikely that you would be able to complete your work without correcting any previous work and, indeed, you would have a duty to draw to the attention of the contractor any defects you discover (**document 3.04.2**).

3.04.1   Letter if contractor tries to pass costs of defects to substitute sub-contractor

This letter is only suitable for use with NAM/SC

To the Contractor
(Copy to the Architect)

Dear Sirs

[Heading]

We are in receipt of your letter dated *[insert date]* by which you appear to be asking us to rectify defects in the existing works carried out by the former named person at our own cost.

We refer you to the conditions of sub-contract which clearly set out the extent and scope of the Works in the numbered documents. Correcting existing defects at our own cost is not part of our obligations and we decline to do so.

[If appropriate, add:]

We are prepared to carry out corrective work if you will confirm your acceptance that we shall be entitled to valuation of such work as a variation in accordance with clause 16.

Yours faithfully

3.04.2   Letter if defects found by substitute sub-contractor
This letter is only suitable for use with NAM/SC

To the Contractor

Dear Sirs

[*Heading*]

Since commencing work on site, we have discovered [*specify details of defects*]. This work was carried out by the previous named person and obviously it will have to be rectified before we continue with the Works we are contracted to do.

In order to avoid delay and disruption, we should be pleased to receive your written directions by [*insert date*].

Yours faithfully

3.05 ESA/1 and design considerations

It is quite likely that you will be expected to undertake some design work as part of the sub-contract works. Indeed, it is probably part of the reason the architect decided that this part of the main contract works should be entrusted to a named person. Under the IFC 84 form of contract, the contractor has no liability for any part of the design. Clause 3.3.7 makes clear that, even when you are given design responsibility, the contractor has no similar liability to the employer. In this respect, the contract and sub-contract are not 'back to back'. Therefore, although the invitation to tender NAM/T section I may ask you to quote for design, although you may quote for design in NAM/T section II, and although the articles of agreement NAM/T section III may show that you have entered into sub-contract on that basis, the contractor will have no need to take action against you and the employer will be unable to take effective action against you for a design defect. Mainly for this reason, the architect will ask you to enter into the RIBA/CASEC Form of Employer/Specialist Agreement ESA/1.

This agreement establishes a direct contractual link between employer and named sub-contractor. The form will be completed in accordance with procedure A or B. Procedure A is used when the architect has sufficient information to enable him to invite a final tender from you. Otherwise procedure B will apply and you will submit an approximate estimate. Under procedure A, a contract comes into effect when the employer signs ESA/1. Under procedure B, ESA/1 acts like a letter of intent, because the employer undertakes to pay your reasonable expenses in the event that you do not enter into the sub-contract.

Paragraph 1 stipulates that you are to comply with any time requirements subject to the employer providing any further information. Paragraph 2 requires you to let the architect have any information you are to provide, in sufficient time to allow the architect to co-ordinate and integrate the sub-contract design into the design of the works as a whole. Paragraph 3 continues the theme and requires you to provide information in good time to enable the invitation of tenders for the main contract works. This is clearly important where you are to be named in the main contract documents.

You have a further obligation to provide information so that the architect can issue his instructions to the contractor at appropriate times. This is to avoid the contractor having any entitlement to extension of time or to additional money in direct loss and/or expense. Strangely, there is no provision within the document if you are in breach of your obligations, and the employer will be thrown on his common law rights to sue for damages for breach. He certainly is not entitled to 'set-off' such amounts from payments to the contractor. If the contractor fails to object and tries to pass on the deduction, you must make the position absolutely clear (**document 3.05.1**).

Paragraph 4 states that the employer is entitled to use your drawings for carrying out and completing the works and for maintaining or altering them. This amounts to a licence. Even without this statement, a similar term would probably be implied provided you had received sufficient payment. Even with this clause in place, you may be able to object to the employer's use of your drawings if you have not been paid, in view of the wording of the clause which refers to information provided 'in accordance with this Agreement' (**document 3.05.2**).

The important part of the agreement is contained in paragraph 5 which puts an obligation on you to use reasonable care and skill in:

- the design of the sub-contract works insofar as you have designed them; and
- the selection of materials and goods insofar as you have selected them; and
- the satisfaction of any performance specification.

The agreement lays down no sanction if you are in breach of this clause and the employer must resort to his common law rights. Undoubtedly, his first move will be to consult the architect who should write and notify you of the problem. **Document 3.05.3** is a possible response.

Paragraphs 7.1 and 7.2 are optional and provide for the employer to require you to proceed with the purchase of materials or goods or the fabrication of components for the sub-contract works. There is also provision for payment and ownership of the goods passing to the employer in the event that a sub-contract is not executed. The employer may not be particularly quick in paying you if the works do not proceed, nor in paying you while you await the execution of the sub-contract. Indeed, on a strict reading of paragraph 7.2, you may not be entitled to any payment until the execution of a sub-contract is formally abandoned. This would be a ridiculous situation (**documents 3.05.4 and 3.05.5**).

3.05.1   Letter if contractor attempts to pass on the employer's set-off
This letter is only suitable for use with NAM/SC
By fax and recorded delivery

To the Contractor
(Copy to the Employer)

Dear Sirs

[*Heading*]

We are in receipt of your letter dated [*insert date*] by which you
inform us that you intend to set-off, from amounts due to us, sums
which the employer has set-off from you in respect of our alleged
breach of the form of Employer/Specialist Agreement ESA/1.

Clearly, the employer is not entitled to set-off from you sums as
damages for alleged breach of a contract to which you are not a
party. In failing to pay you the sum certified, the employer is
himself in breach and you have remedies, both contractual and at
common law, for such breach. Whether or not you choose to
pursue such remedies is a matter for you, but in any event, you
may not set-off such sums from us. In the circumstances, we need
not begin to consider your failure to comply with the set-off
provisions in clause 21.

If, within three working days of the date of this letter, you do not
confirm that you no longer intend to attempt the set-off, we shall
instruct our solicitors to take appropriate action through the courts
on our behalf.

Yours faithfully

Copy: Solicitor

3.05.2   Letter if employer uses drawings and sub-contractor not paid
This letter is only suitable for use with ESA/1

To the Employer

Dear Sirs

[*Heading*]

We understand that you intend to make use of the drawings and
other information we prepared in connection with our work on this
project.

Please take this as formal notice that we have not yet received
full/proper [*delete as appropriate*] payment from the main
contractor and until such time as we do receive such payment, you
have no rights to make use of our drawings and information. Any
attempt so to do will be a breach of our copyright and we shall
instruct our legal advisors to take suitable action on our behalf to
prevent such use.

We believe that we retain such rights under the terms of paragraph
4 of the contract ESA/1 which provides that your entitlement refers
only to drawings and information provided 'in accordance with this
Agreement'. It is our view that our obligations to provide
information under the 'Agreement' cannot be separated from
obligations under the sub-contract, because the one is collateral to
the other.

Yours faithfully

3.05.3 Letter if architect alleges a design problem
This letter is only suitable for use with ESA/1

To the Architect
(Copy to the Employer)

Dear Sirs

[*Heading*]

We are in receipt of your letter dated [*insert date*]. It appears a little premature to allege that we are in breach of our obligations under paragraph 5 of the Agreement ESA/1. Our obligations are set out only insofar as we are responsible for design, choice of materials or satisfaction of a performance specification. The overall responsibility for these items lies with the architect and it has yet to be demonstrated or even implied that the problem to which you refer is the result of any breach on our part.

However, in view of the seriousness of your accusation, we will arrange to visit the site and carry out an inspection. If, as we expect, the matter is not our concern, we shall take your letter as an instruction and we shall bill the employer for the cost of the visit and inspection. We propose to call on site on [*insert date*] at [*insert time*]. Please confirm that it will be convenient and arrange for us to have access to the building.

Yours faithfully

3.05.4   Letter if employer does not pay for goods or fabrication ordered
This letter is only suitable for use with ESA/1
By fax and post

To the Employer
(Copy to the Architect)

Dear Sirs

[*Heading*]

We refer to your letter dated [*insert date*] in which you instructed us to [*insert nature of the supply of goods or fabrication*] under the provisions of paragraph 7 of the Agreement ESA/1. We supplied the goods/completed the work [*delete as appropriate*] by [*insert date*] and although the sub-contract has been executed, we have not yet received any payment. We should be pleased if you would instruct the architect to include the amount in the next certificate and confirm your action by return of post. If we do not receive such confirmation by [*insert date*], we shall instruct our solicitors to recover the money directly from you under ESA/1 as payment for goods supplied/work done [*delete as appropriate*].

Yours faithfully

3.05.5   Letter if employer will not pay and no likelihood of a sub-contract

This letter is only suitable for use with ESA/1

To the Employer
(Copy to the Architect)

Dear Sirs

[*Heading*]

We refer to your letter of the [*insert date*] in which you instructed us to [*insert nature of the supply of goods or fabrication*] under the provisions of paragraph 7 of the Agreement ESA/1. The supply of goods/work [*delete as appropriate*] was completed on the [*insert date*] and at today's date we have received no payment.

Paragraph 7.2 states that if the sub-contract is not entered into, you must pay us for the goods/work [*delete as appropriate*]. We understand that the 21 days stipulated in the main contract has elapsed since its execution and the contractor has not entered into a sub-contract with us. We, therefore, call upon you to pay us forthwith.

If we do not receive your cheques within seven days from the date of this letter we will put the matter into the hands of our lawyers.

Yours faithfully

Chapter 4

Nominated Sub-Contractors

4.01 Introduction

The nomination provisions (clause 35) in JCT 80 are notorious for their length and complexity. This is in no small measure due to the fact that the clause attempts to reconcile the irreconcilable. The employer wishes to be able to instruct the contractor to use a particular sub-contractor for a particular task, but the employer wants no further responsibility for the consequences. This bald premise was found to be unworkable when it was enshrined in JCT 63 short clause 27 as the courts soon pointed out. Consequently, the clause has increased about eightfold in length since JCT 63 and there are continual amendments published to overcome perceived shortcomings.

JCT 80 defines the circumstances in which a sub-contractor is considered to be nominated in clause 35.1. It provides for the procedure for nomination in clauses 35.3 to 35.9. It was revised in March 1991 by Amendment 10 and hence it is commonly referred to as the '1991 procedure', presumably to differentiate it from its predecessors: the 'basic' and 'alternative' methods. The 1991 procedure is supposed to make the nomination procedure simpler.

Clause 35 also deals with extension of the sub-contract period (clause 35.14), failure to complete the sub-contract works (clause 35.15), practical completion of the sub-contract works (clause 35.16), payment (clause 35.17-19) and renomination (clause 35.24).

4.02 The 1991 procedure

The procedure is complex from the architect's point of view, but less so from your standpoint. Apart from a possible initial telephone enquiry, the first you will know about the procedure is when you receive from the architect the Invitation to Tender (NSC/T part 1) which the architect will have completed, together with a blank copy of the Tender (NSC/T part 2) and a copy of the Employer/Nominated Sub-Contractor Form of Agreement (NSC/W). These documents should be examined with care, because it is quite possible that the architect has made a mistake or omitted to complete some part (**document 4.02.1**).

You must complete the tender and sign or execute NSC/W as a deed and return the documents to the architect. At this point, there are two important things to note about your tender. The first is that, under English law, you are entitled to withdraw your tender at any time before it is accepted. That is despite any undertakings to leave your tender open for acceptance for a stated period of time, unless you make such undertakings in return for some consideration from the employer or the contractor (**document 4.02.2**). The architect will not be pleased and you may have to explain the situation (**document 4.02.3**). The second is that the tender makes no provision for you to qualify your tender in any way other than the options set out in the NSC/T part 2 document. If you do so qualify, the architect cannot properly carry out the nomination procedure. However, there may well be situations where you can offer a substantial saving in cost or a more effective method of carrying out the sub-contract work. In this case, you should inform the architect by telephone or fax as soon as possible and he may authorise a re-tender on the basis of your suggestions (**document 4.02.4**). Alternatively, you can submit your tender in the normal way under cover of a letter which sets out your suggested modifications (**document 4.02.5**).

If the architect wishes to nominate you, he will ask the employer to sign your tender as approved and also to execute the agreement NSC/W. The architect will send a copy of NSC/T parts 1 and 2, the numbered tender documents and a copy of the completed NSC/W, together with a nomination instruction on NSC/N to the contractor. You will receive a copy of NSC/N, NSC/W and the completed appendix to the main contract. Once again, you should carefully examine the documents and notify both architect and contractor if there are any discrepancies or if any parts have been amended since you tendered (**document 4.02.6**). At this point, you will be aware of the name of the contractor, perhaps for the first time, and you are entitled to withdraw your tender without giving reasons provided you act within seven days of receiving notice (**document 4.02.7**).

If you do not withdraw, the next stage involves agreeing the matters in the Particular Conditions (NSC/T part 3). If you do not agree, you should notify the contractor immediately (**document 4.02.8**). Once agreed, the Agreement (NSC/A) can be entered into with the contractor. That forms the sub-contract.

If the sub-contract has not been executed within ten days of receipt by the contractor of the architect's nomination, the contractor must notify the architect either of the date when he expects agreement to be reached, or that the failure to agree is due to matters which the contractor must specify. In response, the architect may fix a new date for the agreement to be concluded, or he may issue instructions in regard to the matters in dispute, or he may cancel the nomination, or he may inform the contractor in writing that he does not consider that the matters justify the failure to agree, and the main contract then states in clause 35.9.2 that the contractor 'shall comply' with the nomination instruction. It is extremely doubtful, to say the least, whether the architect has the power to compel the contractor in this way, because the architect is actually instructing the contractor to agree. Such an instruction is not a practical nor a legal possibility. If your views are the cause of the failure to agree, you must be firm in your stance (**document 4.02.9**).

4.02.1   Letter if architect has failed to complete invitation to tender
This letter is only suitable for use with NSC/C

To the Architect

Dear Sirs

[*Heading*]

Thank you for your letter of the [*insert date*] inviting us to tender for [*insert details*] and enclosing NSC/T parts 1 and 2, and a copy of the Employer/Sub-Contractor Agreement NSC/W. We shall be happy to submit a tender.

When we examined NSC/T part 1/NSC/W [*delete as appropriate*], we noticed that there were a number of instances where spaces have been left blank. We have taken a copy of the form and we are preparing our tender on the basis of the information included, but we return the original form herewith. Perhaps you will fill in the places we have flagged and return it to us as soon as possible, bearing in mind the timescale you have set for us.

We hope this is all perfectly clear, but please telephone if we can be of further assistance.

Yours faithfully

4.02.2   Letter if withdrawing offer
This letter is only suitable for use with NSC/C
By fax and post

To the Contractor [if known] and the Employer
(Copy to the Architect)

Dear Sirs

[*Heading*]

We refer to our tender on NSC/T part 2, dated [*insert date*] in the sum of [*insert amount in words and figures*] for the execution of the above work.

Our tender is hereby withdrawn and it is no longer capable of acceptance.

Yours faithfully

4.02.3　Letter if architect questions the withdrawal

To the Architect
(Copy to the Employer and the Contractor)

Dear Sirs

[*Heading*]

Thank you for your letter of the [*insert date*] from which we note that you consider that we are not entitled to withdraw our tender until the period specified in NSC/T, part 2, page 8, item 3, has elapsed.

It is well established that, under English law, we are so entitled unless we have received some consideration for leaving the tender open. That is not the case in this instance. The withdrawal was prompted by purely commercial considerations and we are prepared to discuss the matter with you if that would be helpful. We shall, however, be looking for somewhat improved payment [*amend as appropriate*] terms.

We look forward to hearing from you.

Yours faithfully

4.02.4 Letter if sub-contractor wants to change tender terms
This letter is only suitable for use with NSC/C
By fax and post

To the Architect

Dear Sirs

[*Heading*]

We refer to your invitation to tender on NSC/T part 1 dated [*insert date*]. We shall be delighted to submit a tender as you request, but we confirm our telephone call to your office earlier today when we stated that it appears that we can save a substantial amount of cost/carry out the work in a more effective way [*delete as appropriate*] by [*insert details of the proposal in brief*].

We are continuing to calculate our tender as instructed in the invitation and if you wish us to amend our submission in the light of this letter we should be pleased to receive your agreement by [*insert date*] to avoid abortive tendering work on the wrong basis.

We look forward to hearing from you.

Yours faithfully

4.02.5   Letter with tender, suggesting modifications
This letter is only suitable for use with NSC/C
Registered post/recorded delivery

To the Architect

Dear Sirs

[*Heading*]

We have pleasure in enclosing our tender on NSC/T, part 2, together with the completed Agreement NSC/W.

[*Then, either:*]

In completing the tender we noticed a number of areas where we could suggest improvements and savings in cost as follows: [*list the suggestions briefly*]. If any or all of these suggestions are of interest to you, we should be happy to discuss them in more detail.

[*Or:*]

In completing the tender, we noticed a number of areas where it would be advisable to amend your proposals for [*insert 'safety', 'health', 'structural', etc as appropriate*] reasons as follows: [*list the suggestions briefly*]. We should be pleased to discuss these points in more detail.

Yours faithfully

4.02.6 Letter if amendments on documents at nomination stage
This letter is only suitable for use with NSC/C

To the Architect
(Copy to the Contractor)

Dear Sirs

[Heading]

We are in receipt of your nomination instruction NSC/N dated
[insert date], Agreement NSC/W and the completed appendix to
the main contract.

The documents vary from the information on which we tendered as
follows: *[list changes]*. We are not prepared to enter into a sub-
contract on the revised basis and we return the documents
herewith. If you will reinstate the original terms, we shall be happy
to proceed to the next stage of the nomination process.

Yours faithfully

4.02.7   Letter if sub-contractor withdraws within seven days
This letter is only suitable for use with NSC/C

To the Employer
(Copy to the Architect)

Dear Sirs

[*Heading*]

We are in receipt of your nomination NSC/T dated [*insert date*]
and we see that the main contractor is [*insert name*]. Under the
provisions of paragraph 1.2 of the Agreement NSC/W, we hereby
state that such Agreement shall cease to have effect except as
therein provided and that the offer on Tender NSC/T part 2 is
withdrawn notwithstanding any approval by your signature on page
8 thereof.

Yours faithfully

4.02.8   Letter if sub-contractor does not agree NSC/T part 3
This letter is only suitable for use with NSC/C

To the Contractor
(Copy to the Architect)

Dear Sirs

[Heading]

We are in receipt of the completed NSC/T part 3, to which you are
requesting our signature. We are not prepared to sign the
document as received, because we disagree with a number of the
items as follows: *[list the disagreed items and proposed changes]*.

We, therefore, return the document herewith. If you will amend
these items as noted, we will be prepared to sign.

Yours faithfully

4.02.9   Letter if architect attempts to compel nomination
This letter is only suitable for use with NSC/C

To the Contractor
(Copy to the Architect)

Dear Sirs

[*Heading*]

We have received a copy of an instruction of the architect, dated
[*insert date*] under main contract clause 35.8.2, which purports to
instruct you to comply with the nomination instruction.

We are advised that the architect has no power to instruct you to
reach agreement with us or if he has such power, it is ineffective in
this instance. Moreover, we have satisfied paragraph 1.1 of
Agreement NSC/W in that we have stated our good reasons for
being unable to enter into a sub-contract with you. If you are able
to remove the reasons for our inability, we shall be pleased to
comply.

Yours faithfully

4.03 Other procedures

The only procedure which will result in the nomination of a sub-contractor under clause 35 of JCT 80 is the procedure outlined in section 4.02 above. This is stated in clauses 35.1 and 35.3. If the architect is short of time or fails to appreciate that only the 1991 procedure will result in nomination, he may try to shortcut the system by sending you an informal invitation to tender (perhaps by telephone) or in some other way, he may fail to operate the procedure properly. Because the correct operation of the procedure offers you several safeguards, you must be vigilant in responding to the architect (**documents 4.03.1** and **4.03.2**).

4.03.1   Letter if architect sends informal invitation to tender
This letter is only suitable for use with NSC/C

To the Architect

Dear Sirs

[*Heading*]

Thank you for your letter of the [*insert date*] inviting us to tender
in connection with the above project. We understand from your
letter that we are to be nominated to the main contractor. May we
draw your attention to clauses 35.1 and 35.3 of the conditions of
contract which make clear that all nominations must be effected in
accordance with clause 35.4 to 35.9 inclusive. Any other attempt at
nomination would be ineffective and would result in us entering
into a domestic sub-contract with the main contractor. Since there
are valuable safeguards for us if properly nominated, we must
decline to tender in this instance. If, however, you wish us to
submit a tender in accordance with the proper procedure, we shall
be delighted to comply.

Yours faithfully

4.03.2 Letter if architect does not operate the nomination procedure correctly

This letter is only suitable for use with NSC/C

To the Architect

Dear Sirs

[Heading]

Thank you for your letter/instruction *[delete as appropriate]* of the *[insert date]* in which you *[give details of the letter or instruction]*. May we draw your attention to what we feel sure is an oversight on your part? Clauses 35.4 to 35.9 inclusive of the main contract set out the procedure for nomination. Under the provisions of clauses 35.1 and 35.3, nomination can only be achieved by carrying out the procedure as set down. Because there are valuable safeguards for us if we are correctly nominated, we always insist on the correct procedure being adopted; otherwise we decline nomination.

Yours faithfully

4.04 Main contractor's right of objection

When the contractor receives the architect's nomination instruction NSC/N he has seven days in which to make a reasonable objection if he so wishes (clause 35.5.1). Clause 35.5.2 allows the architect to issue further instructions to remove the cause of the objection. If the contractor objects to you, perhaps because you have been troublesome on other projects, the architect may cancel the nomination, sending you a copy. The feeling may be mutual and the contractor's objection may have crossed the architect's threshold neck and neck with your own objection. If, however, you believe that the contractor's objection is ill-founded, unreasonable or even defamatory, you should waste no time in notifying the architect so as to safeguard future tendering opportunities (**document 4.04.1**).

4.04.1 Letter if contractor unfairly objects to the sub-contractor
This letter is only suitable for use with NSC/C

To the Architect

Dear Sirs

[*Heading*]

We are in receipt of a copy of your instruction number [*insert number*] dated [*insert date*] issued under the provisions of clause 35.5.2 of the main contract nominating another sub-contractor as a result of the main contractor's objection under clause 35.5.1.

You kindly made us aware of the substance of the contractor's objection and we realise that you had no real alternative but to re-nominate. However, we must put on record that the contractor's assertions have no basis in fact. We are prepared, and request the right, to present the facts to you, because we believe that the contractor's assertions are bordering on, if not actually, defamatory.

We would ask you, in any event, to confirm that this contractor's statement will not affect the likelihood of us receiving further tender enquiries from you.

Yours faithfully

4.05   Renomination

Clause 35.24 deals with the situation if renomination becomes necessary. Renomination will become necessary in the following circumstances:

- If the nominated sub-contractor defaults in respect of any of the matters set out in NSC/C clauses:

    7.1.1 Suspending the sub-contract works without reasonable cause;

    7.1.2 Failing to proceed in accordance with clause 2.1 without reasonable cause;

    7.1.3 Fails to remove defective work after notice in writing from the contractor and, as a result, the works are materially affected or if he wrongly fails to rectify defective work;

    7.1.4 Fails to comply with the assignment and sub-letting clauses; and after consultation with the architect, the nominated sub-contractor's employment is determined.

- The nominated sub-contractor's insolvency and the sub-contract is not continued under the supervision of the receiver or manager.

- The nominated sub-contractor determines his employment under NSC/C, clause 7.6, the contractor's default.

- The employer has required the contractor to determine the employment of the nominated sub-contractor under NSC/C, clause 7.3, corruption.

- Work which the nominated sub-contractor has properly carried out has to be re-executed, because the contractor or any other sub-contractor has complied with an instruction of the architect under JCT 80, clauses 7, 8.4, 17.2 or 17.3, and the nominated sub-contractor cannot be required, and will not agree, to carry out the work.

In such cases, the architect will have to go through the 1991 procedure to nominate a substitute sub-contractor. If you are the substitute sub-contractor, you will not be concerned about the history of the project except in the sense that you may be treated in the same fashion as the previous sub-contractor. When tendering, however, there will be a need to take special care, because completing the work of another sub-contractor is notoriously difficult. There may be a particular need to set out your own terms (**document 4.05.1**).

4.05.1   Letter if asked to tender to complete sub-contract work
This letter is only suitable for use with NSC/C

To the Architect

Dear Sirs

[*Heading*]

Thank you for your invitation to tender dated [*insert date*] in
connection with the above project and we confirm receipt of
documents NSC/T parts 1 and 2 and Agreement NSC/W, together
with specification and drawings referred to therein.

We note that we are being asked to tender for the completion of
work commenced, but not completed, by another sub-contractor
and in which there may well be defects. In circumstances such as
these, we are delighted to assist you by tendering, but we insist on
the inclusion of our own particular terms in the sub-contract to take
precedence over the standard terms. A copy of our terms are
attached and we look forward to receiving your agreement before
we commence tendering procedures.

Yours faithfully

4.06   NSC/W and design considerations

Virtually every nominated sub-contract has some measure of design work as an intrinsic part of it. In many instances, sub-contractors are nominated under JCT 80 because that portion of the work is particularly complex. This is often the case with services such as electric installations, computer equipment or mechanical installations. The contractor has no design liability whatsoever to the employer unless the Contractor Designed Portion Supplement is included or clause 42 has been operated. Clause 35.21 entirely excludes the contractor's liability in respect of any design works associated with a nominated sub-contract.

The invitation to tender, NSC/T part 1, may require you to include for design services in your tender, and your agreement with the contractor, NSC/T part 3, may include such services, but the contractor will be unlikely to take any action against you for a design defect, because he will have suffered no loss. The employer has no contractual relationship with you and, therefore, he cannot take effective action. This is the principal reason why you are required to enter into a collateral warranty with the employer, NSC/W.

The agreement establishes a direct contractual link between you and the employer. As is the case with ESA/1, discussed in Chapter 3, the most important provision deals with your design responsibility. You warrant that you have exercised all reasonable skill and care in:

- the design of the sub-contract works insofar as you have designed them; and
- the selection of the kinds of materials and goods insofar as you have selected them; and
- the satisfaction of any performance specification.

There are slight differences in the wording between ESA/1 and NSC/W, but we believe that they amount to much the same in practice. There is no contractual sanction if you are in breach of this term and the employer is thrown back on his common law remedies. The architect will no doubt write to you about any problem. **Document 4.06.1** is a possible reply.

Paragraph 2.2 is intended for the employer to use if he wishes to instruct you to carry out design, to purchase materials or to fabricate components. Paragraph 2.2.2 contains elaborate provisions to enable the employer to pay for design work and subsequently gain a licence to use such work to carry out the sub-contract works, and to enable payment to be made for materials and components upon which ownership passes to the employer. Ownership will not pass, of course, if any of your suppliers have a retention of title clause, until they are paid. Paragraph 2.2.3 deals with payment for work which you carry out as instructed under clause 2.2.1, but which turns out to be abortive for reasons beyond your control. Paragraph 2.2.4 covers your obligation to provide a written statement after nomination, to avoid the danger that the employer will pay twice for the same materials, design or components (**document 4.06.2**).

Paragraph 3 details your obligations to provide information so as not to delay the architect in issuing instructions under the main contract, thus giving the contractor an entitlement to an extension of time or to loss and/or expense. It also gives you an obligation to the employer, to perform in such a manner that the contractor is not entitled to an extension of time under JCT 80 clause 25.4.7 and that the architect will not be obliged to consider the issue of an instruction to determine your employment. Once again, the employer has no contractual remedy if you are in breach, but no doubt the architect will write to you as a first step. Depending upon the precise circumstances, **document 4.06.3** may be suitable. There are some paragraphs which are to your advantage. Paragraph 4 stipulates that the architect will operate clause 35.13.1 which requires him to notify you of the amount included for you in any interim or final payment. If the architect fails to notify you, immediately write to the employer: this is important (**document 4.06.4**).

Paragraph 5.1 obliges the architect to operate the provisions of JCT 80 clauses 35.17 to 35.19. This entitles you to final payment no later than 12 months after the date of practical completion. If the amount is not included in an interim certificate, you must write to the employer (**document 4.06.5**). Paragraph 7.1 places a duty on the architect to operate JCT 80 clause 35.13 with regard to direct payment if the contractor defaults in paying you monies notified to you as certified for your work. This is a very important entitlement and you must pursue it vigorously (**document 4.06.6**). The employer's duty to pay you directly is removed under JCT clause 35.13.5.3.4, if there is a petition for winding up of the main contractor of if there is a resolution to that effect. Paragraph 7.2 requires you to pay back any directly paid monies if the employer furnishes you with reasonable proof that such a petition or such a resolution was in existence when payment was made. You will have to be satisfied before you pay the employer (**document 4.06.7**).

4.06.1   Letter if architect alleges a design problem
This letter is only suitable for use with NSC/C

To the Architect
(Copy to the Employer)

Dear Sirs

[*Heading*]

We are in receipt of your letter of the [*insert date*] in which you allege that we are in breach of our obligations under paragraph 2.1 of the Agreement NSC/W. Our obligations are set out only insofar as we are responsible for design, choice of materials or satisfaction of a performance specification. Your allegation, therefore, appears premature. You have the overall responsibility for these items and it has not been implied, much less demonstrated, that your problem is our liability.

In view of the seriousness of your accusation, we are referring it for the attention of our legal advisors. In the meantime, we propose to call at the site on [*insert date*] at [*insert time*] to inspect. We reserve the right to bill the employer with all the costs we incur as a result of your letter if, as we expect, the problem is not our concern.

Yours faithfully

4.06.2   Written statement from the sub-contractor under NSC/W
This letter is only suitable for use with NSC/C

To the Employer [in duplicate]
(Copy to the Architect and to the Contractor)

Dear Sirs

[*Heading*]

Please take this as the written statement required of us by paragraph 2.2.4 of Agreement NSC/W.

We confirm that the amount paid to us under paragraph 2.2.2 should be credited to the main contractor and that we will allow such credit in the discharge of any amount due in respect of the sub-contract works.

Yours faithfully

4.06.3 Letter if architect alleges sub-contractor has caused the contractor to become entitled to an extension of time

This letter is only suitable for use with NSC/C

To the Architect
(Copy to the Employer)

Dear Sirs

[*Heading*]

We are in receipt of your letter dated [*insert date*] in which you allege that we are in breach of our obligations under paragraph 3 of the Agreement NSC/W. In order to be able to recover any monies from us, the employer must demonstrate that the contractor was validly entitled to an extension of time and that such entitlement resulted directly from our default. The employer has not so demonstrated either or both of those propositions and we reject your allegations. Any claim you may bring against us in this respect will be robustly defended.

Yours faithfully

4.06.4 Letter if architect fails to notify amounts included in his certificate

This letter is only suitable for use with NSC/C

To the Employer
(Copy to the Architect)

Dear Sirs

[*Heading*]

Please take this as formal notice that the architect has failed to operate clause 35.13.1 of the main contract in respect of the last certificate. This is in breach of paragraph 4 of the Agreement NSC/W and we should be grateful if you would instruct the architect to operate clause 35.13.1 forthwith.

Yours faithfully

4.06.5   Letter if architect fails to operate the final payment provisions
This letter is only suitable for use with NSC/C

To the Employer
(Copy to the Architect)

Dear Sirs

[*Heading*]

Please take this as formal notice that the architect has failed to
operate the provisions of clauses 35.17 of the main contract in
respect of the final payment. This is in breach of paragraph 5.1 of
Agreement NSC/W and we should be grateful if you would instruct
the architect to operate clause 35.17 forthwith.

Yours faithfully

4.06.6 Letter if the architect fails to operate the direct payment procedure

This letter is only suitable for use with NSC/C

To the Employer
(Copy to the Architect)

Dear Sirs

[*Heading*]

Please take this as formal notice that the architect has failed to operate the provisions of clause 35.13 of the main contract. This is despite the clear failure of the contractor to provide proof of payment as required by clause 35.13.3 and our letter to the architect on the subject dated [*insert date*], a copy of which is enclosed.

This is a breach of paragraph 7.1 of the Agreement NSC/W and we should be grateful if you would instruct the architect to operate clause 35.13 forthwith. Failure to do so will result in immediate legal action.

Yours faithfully

4.06.7   Letter if employer requests repayment on main contractor's insolvency

This letter is only suitable for use with NSC/C

To the Employer
(Copy to the Architect)

Dear Sirs

[*Heading*]

We are in receipt of your letter dated [*insert date*] in which you request repayment of the sum of [*insert amount in words and figures*] paid to us under the provisions of clause 35.13 of the main contract.

You state that you are applying for repayment under paragraph 7.2 of the Agreement NSC/W and we, therefore, require reasonable proof from you that at the time of such payment there was in existence a petition or resolution to which clause 35.13.5.4.4 refers.

Yours faithfully

Chapter 5

Commencement and Progress

5.01 Programme

The programme is not usually a contract document. Certainly, it is not intended to be a contract document under any of the standard forms of sub-contract. Very often, however, you will receive a copy of the main contractor's programme as one of the tender documents. You may also be required to submit your proposed programme with your tender for the sub-contract works. If not, you will be requested to submit your programme before you start work.

If your programme is submitted with your tender and it is subsequently bound into the contract documents, or if the formal documentation is never completed but your tender is clearly based on the programme as submitted and the contractor accepts the tender as it stands, it may become a contract document. If that were so, and you fail to comply with your own programme, it would be a breach of contract for which, theoretically, the contractor could recover any losses he could prove. Conversely, the contractor might require you to carry out the work in a different sequence. In those circumstances, the contractor's requirement would amount to a variation for which you would be entitled to be paid and you should respond to the contractor's directions accordingly (**document 5.01.1**). If your method statement rather than a drawn programme is a contract document, the same situation results.

If it is not a contract document, what value or force can be attached to your programme? The answer is that it will be very persuasive evidence of the way in which you intended to carry out the work. It could be vital in any attempt to recover loss and/or expense under the sub-contract provisions.

Your contract will set out the period during which you must carry out the work. Both NSC/C and NAM/SC do this by reference to the period in the form of tender. Indeed, NAM/SC requires you, in clause 12.1, to carry out and complete the sub-contract works in accordance with (a) the period inserted by the architect within which the sub-contract works must be carried out and (b) the period you have inserted for execution of the works. Needless to say, these periods may be different and this is something which must be clarified before you enter into any sub-contract. Failure to do so, however, may act in your favour if the contractor alleges you are in delay (see section 7.06).

DOM/1 and DOM/2 simply refer to the details in the Appendix Part 4. In all cases, you are required to carry out the work 'reasonably in accordance with the progress of the Works', but subject to receipt of notice to commence work on site. The result of that wording appears to be that the actual start date on site is subject to change for which you are not entitled to any recompense. This is, of course, always subject to the delay being reasonable in the context of the contract as a whole. If it could be argued that the delay in commencing the sub-contract works is such that it falls totally outside the stipulated commencement period, it is unlikely that you would be bound even to start the works and you should respond to such a notice forcibly (**document 5.01.2**).

You must also be prepared to work along with the contractor. You must, therefore, allow for the fact that the works may not proceed with textbook smoothness. Serious disruptions are dealt with by the extension of time provisions (see Chapter 7).

The extent to which a sub-contractor must organize his work to fit in with the main contractor is often under-estimated. If the terms of the sub-contract require you to carry out the sub-contract work to the contractor's requirements or to his instructions or similar words, and there is no other indication of a programme for the works, your obligation will be to work as and when directed by the contractor. You will have no claim for so doing even if the work is very much disrupted or protracted (see Chapter 8). You will often find that some such words have been inserted rather than specific dates or periods of time in the contract documents. You must resist all such attempts to dispense with a precise sub-contract period (**document 5.01.3**).

It is always a good idea to submit your programme for 'approval' by the contractor (**document 5.01.4**). Although the contractor's approval does not give the programme contractual force if it is not in fact a contract document, once the contractor has given his approval, it becomes very difficult for him to argue later that it is unrealistic. This can be very useful in a claims situation.

Programmes are usually submitted in the form of a Gantt or bar chart. Although such programmes are easily understood, it is well worth preparing your programme in the form of a precedence diagram or network analysis. Dependencies of activities can be established and the effect of events upon critical activities can be demonstrated. With the aid of a computerized programme, it is possible to carry out exercises showing the effect on the proposed resources and the result of increasing or decreasing those resources in response to the contractor's directions.

5.01.1   Letter if variation of contract programme ordered

To the Contractor

Dear Sir,

[Heading]

Thank you for your letter dated *[insert date]* in which you
directed us to *[specify the precise details of the change, for
example: 'finish all pipework connections in block E before
proceeding with the remainder of the hot water installation']*. The
contract programme, on which our price is based, clearly envisages
*[specify the original intention, for example: 'that work on the
connections would start in block A, moving on to blocks B, C and D
before block E']*.

We will, of course, comply with your directions for which we will
expect payment under the provisions of clause 16/17 *[substitute
'16' when using NAM/SC or '4.4/4.10' when using NSC/C]*.   Take
this as notice of delay as required by clause 11.2 *[substitute '12.2'
when using NAM/SC or '2.2' when using NSC/C]* and we consider
the matter referred to above to be a relevant event under clause
11.10.5.1 *[substitute'12.7.5.1' when using NAM/SC or '2.6.5.1' when
using NSC/C]*.

[continued]

5.01.1 *contd*

We will furnish you with our estimate of delay in the completion of the sub-contract works and further particulars as soon as possible. We hereby make application under clause 13.1 [*substitute '14.1' when using NAM/SC or '4.38.1' when using NSC/C*] that we are likely to incur direct loss and/or expense including financing charges due to the regular progress of the sub-contract works being materially affected by the matter referred to above which is a relevant matter in clause 13.3.7 [*substitute '14.2.6' when using NAM/SC or '4.38.2.7' when using NSC/C*]. We will submit appropriate information in support of this application in due course.

Yours faithfully

5.01.2   Letter in response to late notice to commence work

To the Contractor

Dear Sir

[Heading]

We are in receipt of what purports to be a notice to commence work dated *[insert date]*.

Your notice is well outside the period during which the sub-contract stipulates a start will be made on site and in this connection we refer you to the Appendix Part 4 *[substitute 'NAM/T, Section 1, item 15' when using NAM/SC or 'NSC/T, Part 3, item 1(2)' when using NSC/C]*.

The date on which we are entitled to commence on site is an important term of the sub-contract. Your failure to give notice in due time is, therefore, a serious breach of the sub-contract and we reserve all our rights and remedies in this matter. Without prejudice to the foregoing, we suggest that a meeting would be useful and look forward to hearing from you.

Yours faithfully

5.01.3   Letter regarding imprecise sub-contract period

To the Contractor

Dear Sirs,

[*Heading*]

We are in receipt of your letter of the [*insert date*] with which you enclosed the sub-contract documents for us to sign/seal/execute as a deed [*delete as appropriate*].

These documents are not consistent with the tender documents on which our tender is based. In particular, there is no sub-contract period shown, our obligation being expressed as [*insert the exact wording, e.g., 'To the instructions of the contractor'*]. We are not prepared to enter into a contract on this basis.

We, therefore, return the documents herewith and we look forward to receiving the corrected documents as soon as possible.

Yours faithfully

5.01.4 Letter enclosing programme

To the Contractor

Dear Sirs

[*Heading*]

We have pleasure in enclosing two copies of our programme for the above sub-contract works. You will note that it is presented as a Gantt chart and also in the form of a precedence diagram.

We should be pleased to receive your approval as soon as possible.

Yours faithfully

5.02 Meetings

The world seems to revolve around meetings, and sub-contracting is no exception. It is still customary for the architect to hold site meetings at regular intervals, although some architects now only call such meetings as and when required. In any event, it is unlikely that sub-contractors will be required at the architect's site meetings which normally involve only the consultants, the clerk of works and the contractor. Sometimes, a nominated sub-contractor with design responsibility may be requested to attend. Generally, the contractor is expected to call his own meeting of sub-contractors and suppliers the day before the main contract site meeting. A typical agenda for such a meeting is shown as **document 5.02.1**. The contractor should circulate the agenda a few days before each meeting, but in practice, you will usually get your first sight of it, if there is an agenda, when the meeting begins.

Minutes of each meeting should be taken by the contractor. If there are no formal minutes, which is rare, you should take your own and confirm all the important points to the contractor and any other interested party as soon as possible after the meeting. It is more likely that the contractor will take the minutes and distribute them within a couple of days of the meeting.

When you receive your copy, you should read it carefully. Sometimes what is not recorded is more important than what is recorded. The minute of a meeting is an important document in a claims situation, whether the claim is by you or against you. For this reason alone, you must write to the contractor if there are any inaccuracies or omissions (**document 5.02.2**). It is best to write immediately. You should not wait until the next meeting.

If the minutes of a meeting are recorded at a subsequent meeting as agreed by all present, they will stand as a true record. If there is no agreement, they are merely the contractor's note of what was said. They may still be persuasive evidence unless refuted, which is why it is so important that you respond to every inexactitude. The standard forms require all instructions or directions given by the contractor to be given in writing (clauses 4.2.1 and 4.2.2 of DOM/1 and DOM/2, 5.2.1 and 5.2.2 of NAM/SC and 3.3 of NSC/C). It is not thought that to record an instruction or direction in the minutes of a meeting satisfies this requirement. If you receive minutes which purport to contain directions, you should request proper directions immediately (**document 5.02.3**).

5.02.1   Typical agenda for site meeting

Job Title:
Location:
Ref No:

Meeting No:

1. Minutes of the last meeting.
2. Matters arising and action taken.
3. Weather report and manhours lost.
4. Labour force on site, by trades.
5. Progress reports:
    Main Contractor
    Nominated Sub-Contractors
    Domestic Sub-Contractors
6. Revisions to drawings and variations issued since the last meeting.
7. Workmanship and materials queries.
8. Any other business.
9. Date of the next meeting.

Distribution:
  Nominated Sub-Contractors
  Nominated Suppliers
  Domestic Sub-Contractors
  Selected Suppliers

5.02.2 Letter regarding minutes of the last meeting

To the Contractor
(Copies to all present at the meeting and included in the original circulation)

Dear Sirs,

[*Heading*]

We have examined the minutes of the meeting held on the [*insert date*] which we received today. We have the following comments to make:

[*List comments*]

Please arrange to have these comments published at the next meeting and inserted in the appropriate place in the minutes.

Yours faithfully

5.02.3   Letter if minutes contain directions

To the Contractor

Dear Sirs,

[*Heading*]

We have today received your minutes of the meeting held on [*insert date*]. We note that item [*insert number*] purports to be a direction requiring us to [*insert details*]. Clauses 4.2.1 and 4.2.2 [*substitute '5.2.1 and 5.2.2' when using NAM/SC or '3.3.1' when using NSC/C*] of the sub-contract empower you to issue directions in writing regarding the sub-contract works.

In our view, the item in the minutes cannot reasonably be considered to be a direction properly given under the terms of the sub-contract. For the avoidance of doubt, therefore, we request you to issue us with a clear written direction in regard to the matters set out above, on receipt of which we will comply forthwith.

Yours faithfully

5.03 Discrepancies

Problems commonly arise on site because it is found that the contract documents do not all agree. The standard forms provide for this situation. DOM/1 and DOM/2, clause 2.3, state that if the sub-contractor finds any discrepancy in or divergence between the sub-contract documents and the contractor's directions, he must immediately give a written notice to the contractor specifying the discrepancy or divergence. **Document 5.03.1** is an example of such a notice. The onus is then upon the contractor to issue directions. In many instances, the direction will constitute a variation to be valued under clause 16. It may give rise to an extension of time under clause 11.10.5.1 and enable you to make application for direct loss and/or expense under clause 13.3.3.

Not every discrepancy will fall under the provisions of clause 2.3. Clause 2.2 helpfully sets out an order of priority of documents so that only if there remains an unresolved discrepancy will clause 2.3 apply. The order of priority can best be set out as follows:

- The appendix prevails over the sub-contract conditions.
- Both documents (known as DOM/1 or DOM/2 as appropriate) prevail over the numbered documents.
- These documents (known as the sub-contract documents) prevail over the terms of the main contract.

By clause 4.1.4, you are also to note any discrepancy you may find in or between the documents listed in main contract clause 2.3 and the numbered documents. Your opportunities to examine the main contract drawings will be limited, but it should be noted that the contractor's directions under this clause may entitle you to an extension of time and enable you to make application for direct loss and/or expense quite apart from their valuation as variations under clause 16.

It is now established that the discrepancy clause in the main contract does not oblige the contractor to search for discrepancies. It is suggested that clauses 2.3 and 4.1.4 of DOM/1 and DOM/2 must be interpreted in the same way so that your duty is simply to notify the contractor if you find a discrepancy. The discrepancy may not become obvious until just before the discrepant item is due to be carried out or even after it has been completed. The contractor will no doubt argue that you should have spotted the problem earlier. You must strongly resist such arguments (**document 5.03.2**).

NAM/SC clause 2.4 does not expressly require you to notify the contractor if you find an inconsistency in or between the sub-contract documents or drawings or the further documents which the contractor must provide to enable you to carry out and complete the works. Instead it places the duty to issue directions for the correction of inconsistencies firmly on the contractor. It is not, however, suggested that if you discover an inconsistency, you should conceal the fact. The contractor would be justified in considering that you were not co-operating to assist in the completion of the project in such circumstances. You should resist any allegation from the contractor that you are liable if a discrepant item is built and the previous letter is applicable. The order of priority of documents is dealt with in clause 2.2 and it can be expressed as follows:

- NAM/T prevails over NAM/SC.
- Both documents prevail over the numbered documents.
- These documents (known as the sub-contract documents) prevail over the terms of the main contract.

NSC/C, clause 1.8, requires you to give written notice to the contractor if you find any discrepancy in or between the documents listed in clause 2.3 of the main contract. That is the equivalent discrepancy clause relating to the contractor. The previous comments regarding discrepancies also apply here and it should be noted that you are likely to have the opportunity of examining a restricted number of main contract documents, probably only those having some reference to the sub-contract works. Once you have given the contractor a notice, he must send it to the architect, requesting instructions under clause 2.3 of the main contract. Any such instructions which affect the sub-contract works must be issued to you by the contractor under clause 3.3.1.

Unlike the other two forms, the main contract terms have priority under NSC/C. The order of priority is as follows:

- Main contract terms prevail over NSC/C.
- These documents prevail over any other sub-contract documents.

5.03.1 Letter if discrepancy found
This letter is not suitable for use with NAM/SC

To the Contractor

Dear Sirs

[*Heading*]

In accordance with clause 2.3 [*substitute '1.8' when using NSC/C*] we bring to your attention the following discrepancies which we have discovered:

[*List, giving precise references of bills of quantities or specification, drawings, etc*]

[*Add, if applicable:*]

In order to avoid delay or disruption to our progress, we require your instructions by [*insert date*].

Yours faithfully

5.03.2 Letter if contractor asserts that discrepancy should have been notified earlier

To the Contractor

Dear Sirs,

[*Heading*]

Thank you for your letter of the [*insert date*]. We did not find the discrepancy to which you refer until [*insert date*]. Our obligation under clause 2.3 [*substitute '1.8' when using NSC/C*] is to notify you of any discrepancy or divergence we find. [*When using NAM/SC substitute for the last sentence: 'We have no contractual obligation to notify you of inconsistencies; it is your obligation under clause 2.4 to issue directions regarding their correction.'*]

Moreover, it is now well established that we have no obligation to check the information you supply. On the contrary, it is your duty to see that such information is correct.

We shall be writing to you within the next few days when the situation on site is clearer. At that time, we will inform you of the expected delay and its effect on the sub-contract completion date. If appropriate, we will also make application in respect of loss and/or expense resulting from the discrepancy.

Yours faithfully

5.04 Directions

Under the terms of DOM/1 or DOM/2, clause 4.2.1, the contractor's power to give you directions is apparently very broad. He may issue 'any reasonable direction' in regard to the sub-contract works. In practice, it is likely that the clause will be given a very tight interpretation so that only directions which may be considered necessary for the proper carrying out and completion of the works, or directions which may be implied from the terms of the contract, will be empowered. An example of the latter would probably be if the contractor gave you directions to postpone the work for a short period or to open up the works for examination. The ordering of variations is expressly empowered.

In addition, clause 4.2.2 stipulates that any written instruction of the architect under the main contract which affects the sub-contract works will be deemed to be a direction if issued to you by the contractor. Although there is no special mechanism set out for challenging the validity of any direction, you are entitled to make reasonable objection if the direction is not reasonable (**document 5.04.1**). If you comply with a direction which the contractor is not empowered to give under the terms of DOM/1 or DOM/2, the result may be that:

- you are not entitled to payment; or
- you are not entitled to payment and you must return the situation to the state which existed before the direction was given; or
- the contractor may be liable to reimburse you.

You may also make reasonable objection if the direction is a variation relating to:

- Access to the site.
- Limitation of working space.
- Limitation of working hours.
- Sequence of working.

Document 5.04.2 is suitable. Although clause 4.2.4 stipulates that you must comply forthwith (i.e. as soon as you reasonably can), this obligation is removed to the extent that you make a reasonable objection. The contractor has power to issue a seven-day compliance notice if he does not think that you are getting on with a direction as expeditiously as he would like. If you do not begin to comply within the seven days, the contractor may employ others to carry out the work and deduct from you all costs incurred. On receipt of such a notice, therefore, you should take some action. If, nevertheless, the contractor proceeds to employ others, a letter is indicated (**document 5.04.3**). It may be that you cannot comply with the instruction for a very good reason. If that is the case, you must put the matter on record so that, in due course, you can resist the deduction of money (**document 5.04.4**).

Although the contractor's directions are stated to be 'in writing', clause 4.4 provides for the situation if a direction is issued by other means, i.e. orally. You are to confirm such a direction within seven days (**document 5.04.5**). If the contractor does not dissent within a further seven days, the direction takes effect from the expiry of that second seven-day period. If the contractor confirms himself within the seven-day period, your obligation to confirm is removed. If you carry out a direction without any confirmation on either side, the contractor may confirm at any time up to the date of the final payment. If the contractor fails to so confirm, not only will you not receive payment, you may also be required to put back the work into the condition it was before the unconfirmed direction was carried out. This is a common cause of dispute easily avoided by sending the appropriate confirmation.

The position under NAM/SC, clauses 5.2 to 5.4 is very similar, with one marked exception. The terms of the main contract (IFC 84) are reflected in that there is no provision for the contractor to give or confirm oral directions. The result of this is that no confirmation on your part can make an oral direction valid. It is probable that if the contractor confirms an oral direction himself, it will, in effect, be as if he issued a written direction on the date he sent his confirmation. If you do receive an oral direction, it is important that you waste no time in requesting the direction in writing (**document 5.04.6**). Do not take any action on the direction in the meantime.

There is extensive provision in clause 5.6 which virtually mirrors the terms of the main contract in the case of work not being in accordance with the sub-contract. In such a case, you are required to make a written statement to the contractor setting out your proposals to establish that there is no similar failure in other parts of your work or in materials or goods supplied. The contractor may issue directions requiring you to open up or test any part of the work if:

- the contractor does not receive your statement within five days of discovery of the failure; or
- your proposal is not satisfactory; or
- safety considerations prevent the contractor from waiting for your proposals.

You are to comply forthwith, but you have the right to object within seven days of receipt of the contractor's directions (**document 5.04.7**). If the contractor does not withdraw or modify his direction to deal with your objection within a further seven days, the matter is automatically referred to arbitration to decide whether the direction was reasonable in all the circumstances. You are required to comply with the contractor's directions at no extra cost except to the extent that the arbitrator decides that any direction is not fair and reasonable.

NSC/C deals with directions of the contractor in clauses 3.3 and 3.10. The position is very similar to that under DOM/1 and DOM/2, with two important differences. The first is that, on receipt of what purports to be an architect's instruction issued to you by the contractor, there is provision under clause 3.11 for you to require the contractor to request the architect to specify the empowering main contract (JCT 80) clause (**document 5.04.8**). The contractor must send you the architect's reply. This is a valuable protection for you because if you carry out the instruction thereafter, it is deemed authorised by the clause specified and you are then entitled to all the benefits such as extensions of time, reimbursement of loss and/or expense and payment for the work done. This is the case even if the architect is wrong and the empowering clause he names does not authorise the instruction. Alternatively, you can refuse to accept the instruction and seek arbitration on the point. Depending on the circumstances, it may be sensible to carry out the instruction while awaiting arbitration, but preserving your rights by writing an appropriate letter (**document 5.04.9**).

Sometimes the situation arises in which the contractor insists that certain work is included in your sub-contract, but which you are equally certain should be an extra. The contractor may refuse to authorise a variation. If the amount of work is substantial and you are confident of your position, you may be able to refuse to carry out the work unless a variation is authorised and, failing such authorisation, to treat the contract as repudiated and sue for damages, although you should not do that without taking appropriate legal advice. Unless your case is cast iron, this is a dangerous course to take because, if you fail, you face the possibility of huge damages. On the other hand if you proceed with the work and attempt to claim later, it may be considered that you have carried out the work on the basis of the contractor's interpretation of the sub-contract and you may not be entitled to any extra payment.

The matter is best settled by arbitration and it should be possible for you to proceed on a 'without prejudice' basis while the dispute is settled in this way, rather than have the whole of the works grind to a halt. **Document 5.04.10** is a suitable letter.

If you carry out work which is not in accordance with the sub-contract and the architect operates the provisions of clause 8.4 of the main contract, clauses 3.5 to 3.9 apply. If the architect consults with the contractor about the problem, the contractor is obliged to consult with you and report back to the architect. The architect has a number of options. He may:

- Instruct the removal of such work from site.
- Allow such work to remain and an appropriate deduction will be made from the sub-contract sum.
- Issue an instruction requiring variations as are reasonably necessary as a result of allowing such work to remain or instructing its removal from site. No payment is to be made and no extensions of time are applicable.
- Instruct such opening up for inspection as is reasonable to establish that there is no similar problem in other parts of the work. To the extent that the instruction is reasonable, no payment is to be made, but extension of time provisions are applicable unless the opening up reveals work not in accordance with the contract.

There is an important provision in clause 3.8 which provides for the situation if defective work by the contractor or other nominated sub-contractors results in some of your work having to be taken down and reconstructed. You are obliged to carry out the taking down or reconstruction of your own work if the instruction to do so is issued to you before the date of practical completion of your sub-contract works. You are entitled to payment on the basis of a fair valuation, reimbursement of loss and/or expense under clauses 4.38 and 4.39, and extension of the sub-contract period under clauses 2.2 to 2.6 where appropriate. If you are given such an instruction, you should question its validity under the provisions of clause 3.11 and confirm that you are entitled to be reimbursed accordingly (**document 5.04.11**).

5.04.1 Letter regarding the validity of a direction
This letter is not suitable for use with NSC/C

To the Contractor

Dear Sirs,

[*Heading*]

We are in receipt of a communication from you dated [*insert date*] which purports to be a direction requiring us to [*insert details*]. We consider that such a direction is not reasonable and not empowered under the terms of the sub-contract because [*insert details of your objection*]. Therefore, take this as reasonable objection [*insert 'under clause 4.3' when using DOM/1 or DOM/2*] to compliance with such instruction.

We look forward to hearing that you withdraw such direction or modify it to meet our reasonable objections.

Yours faithfully

5.04.2   Letter if variation within meaning of 'variation'

To the Contractor

Dear Sirs,

[Heading]

We are in receipt of your direction dated *[insert date]*.   Under the
provisions of clause 4.3 *[substitute '5.3' when using NAM/SC or
'3.3.2' when using NSC/C]* if the direction is a 'variation' within
the definition in clause 1.3, *[substitute '1.4' when using NSC/C]*
we are not obliged to comply to the extent that we make
reasonable objection in writing.

We do hereby object on the grounds that *[specify the objection]*.
We should be pleased, therefore, if you would withdraw the
direction or modify it to overcome this problem. We are taking no
action on this direction until we receive your further directions as
requested.

Yours faithfully

5.04.3 If the contractor proceeds to employ others after seven-day compliance notice issued

---

To the Contractor

Dear Sirs,

[Heading]

Following your letter of the *[insert date]*, you have begun to employ others to carry out the work indicated on your direction number *[insert number]* dated *[insert date]*. This is despite our letter of the *[insert date]* in which we explained the reasons for the delay and outlined our proposals for carrying out the direction. In our view your action is a serious breach of contract for which we will seek appropriate remedies.

Without prejudice to the foregoing, if you will instruct the other persons to withdraw from site forthwith, we will not pursue our strict legal rights on this occasion.

Yours faithfully

5.04.4   Letter on receipt of seven-day notice requiring compliance
Registered post/recorded delivery

To the Contractor

Dear Sirs,

[*Heading*]

We have today received your notice dated [*insert date*] which you
purport to issue under the provisions of clause 4.5 [*substitute '5.4'*
when using NAM/SC or '3.10' when using NSC/C] of the
conditions of sub-contract.

It is not reasonably practicable to comply as you require within the
period you specify because [*insert reasons*]. You may be assured
that we have not forgotten our obligations in this matter and we
intend to carry out your direction number [*insert number*]   dated
[*insert date*] as soon as [*indicate activity*]   is complete. In the
light of this explanation, we should be pleased to hear, by return,
that you withdraw your notice requiring compliance. If we do not
have your reply by [*insert date*] we will indeed immediately
comply, but take this as notice that such immediate compliance will
give grounds for substantial claims for extensions of time and loss
and/or expense.

Yours faithfully

5.04.5   Letter confirming an oral direction
This letter is not suitable for use with NAM/SC

To the Contractor

Dear Sirs,

[*Heading*]

We hereby confirm that on [*insert date*], you orally directed us to [*insert the direction in detail*].   Under the provisions of clause 4.4 [*substitute '3.3.3' when using NSC/C*] you have seven days from receipt within which to dissent in writing.

Yours faithfully

5.04.6   Letter requesting written direction
This letter is only suitable for use with NAM/SC

To the Contractor

Dear Sirs,

[Heading]

On *[insert date]* you orally directed us to *[insert details of the direction]*. In accordance with clauses 5.2.1 and 5.2.2 of the conditions sub-contract, all directions must be in writing. Directions issued other than in writing are invalid.

If you will issue the above direction in writing, we will be delighted to comply forthwith.

Yours faithfully

5.04.7 Letter objecting to compliance with clause 5.6.2 direction
This letter is only suitable for use with NAM/SC

To the Contractor

Dear Sirs,

[Heading]

We are in receipt of your direction number *[insert number]* of the *[insert date]* directing us to *[insert nature of work]* which you purport to issue under clause 5.6.2 of the conditions of sub-contract. We consider such direction unreasonable because *[state reasons]*.

If within seven days of receipt of this letter you do not in writing withdraw the direction or modify it to remove our objection, a dispute or difference will exist as to whether the nature or extent of opening up/testing *[delete as appropriate]* in your direction is reasonable in all the circumstances; such dispute or difference to be referred to immediate arbitration. In such event, we will comply with our obligations pending the result of such arbitration and award of additional costs and extension of time.

Yours faithfully

5.04.8   Letter requiring the specification of empowering main contract clause

This letter is only suitable for use with NSC/C

Dear Sirs,

[*Heading*]

We have received today what purports to be an instruction of the architect dated [*insert date*] requiring us to [*insert substance of the instruction*]. It is our view that the architect is not authorised to issue such instruction.

We require you therefore, in accordance with clause 4.6 of the conditions of sub-contract, to request the architect to specify in writing the provision of the main contract which empowers the issue of the above instruction.

Yours faithfully

5.04.9   Letter if carrying out instruction while awaiting arbitration
This letter is only suitable for use with NSC/4
Registered post/recorded delivery
Without prejudice

To the Contractor

Dear Sirs,

[*Heading*]

We acknowledge receipt of your letter of the [*insert date*] with which you enclosed a copy of the architect's letter of the [*insert date*] purporting to specify the empowering clause for his instruction number [*insert number*] dated [*insert date*].

We are writing separately to you requiring arbitration on the matter. Notwithstanding our requirement for arbitration, we are prepared to agree to your request to execute the work contained in the purported instruction on the clear acceptance by the parties to the main contract that such execution will not prejudice our rights and remedies at arbitration or elsewhere. To that end we should be pleased to receive such acceptance from you and separately from the employer under the main contract, together with an undertaking that you will indemnify us against any loss, expense, costs, claims or proceedings, howsoever arising, as a result of the execution of the above mentioned work

Yours faithfully

5.04.10   Letter if disagreement over whether work is a variation or included in the sub-contract

Registered post/recorded delivery
Without prejudice

To the Contractor

Dear Sirs,

[*Heading*]

We refer to your letters of the [*insert dates*] and ours of the [*insert dates*] relating to [*describe the work*].   It is our firm view that this work is not included in the contract and, therefore, constitutes a variation for which we are entitled to payment.

We are advised that we can refuse further performance until you authorise a variation. If you continue to refuse to so authorise, we may treat the contract as repudiated and sue for damages.

Without prejudice to our rights, we are prepared to carry out the work, leaving this matter in abeyance for future determination by arbitration, if you will agree in writing and confirm that you will not deny our entitlement to payment in such reference if, on the true construction of the sub-contract, the work is held to be not included.

Yours faithfully

5.04.11   Letter if a clause 3.8.1 direction issued
This letter is only suitable for use with NSC/C

To the Contractor

Dear Sirs,

[*Heading*]

We are in receipt of your directions dated [*insert date*] requiring us to [*describe the work*].

Please confirm that such direction is validly issued under the provisions of clause 3.8.1, that we will be reimbursed according to clause 3.8.2 and further that the provisions of clauses 4.38, 4.39, and 2.2 to 2.6 will apply.

Yours faithfully

5.05   Information

There is no specific provision in DOM/1 or DOM/2 for you to receive information from the contractor other than directions. Directions can, of course, include drawings and other information provided that they are clearly indicated as directions. However, if you wish to preserve your entitlement to extension of time and loss and/or expense if necessary information is not received on time, you must specifically apply for it in writing neither too early nor too late having regard to the completion date (clauses 11.10.6 and 13.3.1). It is now established that you can fulfil that obligation by marking your requirements on the programme for the works which you give to the contractor, provided that you keep the programme updated, although it is best to make specific applications by letter. Indeed, you will preserve your entitlement even if you do not apply for the information, provided that the contractor makes application to the architect. It is less open to dispute, however, if you make separate written application (**document 5.05.1**).

NAM/SC, clause 2.3, requires the contractor to provide you with two copies of such further drawings or details as are reasonably necessary to enable you to carry out and complete the sub-contract works in accordance with the sub-contract documents. A very important part of the sub-contract documents is the sub-contract period. The contractor's obligation, therefore, is to provide the information at appropriate times to enable you to finish the sub-contract works within the sub-contract period. The contractor's duty does not depend upon any application from you. If he fails to provide the information, he is in breach of contract. However, your entitlement to extension of time or reimbursement of loss and/or expense under the terms of the sub-contract depends upon your application as described for DOM/1. The appropriate clauses are 12.7.6 and 14.2.1.

Perhaps surprisingly, the position under NSC/C is similar to DOM/1. There is no specific provision for the contractor or the architect to provide drawings, but your entitlement to extension of time and reimbursement of loss and/or expense depends upon your application. Clauses 2.6.6, 4.38.2.1 are applicable.

5.05.1 Letter applying for information

To the Contractor

Dear Sirs,

[Heading]

Take this letter as specific application in writing for the following information which we require in order to carry out the sub-contract works:

[List the information required and the date by which it is required e.g.:

Information required	*Date by which required*
Positions of holes in first floor	*24 October 1990*
Plant positions on roof	*7 November 1990]*

Yours faithfully

5.06   Variations

Variations are basically empowered in DOM/1, DOM/2 and NAM/SC by the clauses dealing with directions. In each case, a variation is defined in clause 1.3. Where NSC/C is used, variations are controlled by instructions of the architect under clause 3.3.1 and the definition is found in clause 1.4. A variation may relate to the alteration or modification of the design quality or quantity of the sub-contract works and it may include the addition, omission or substitution of work, alteration in kinds or standards of materials, removal of work or materials from site and change in any obligations or restrictions with regard to site access, working space, working hours or sequence of working.

If the variation clause did not exist, neither the architect nor the contractor would have the power to instruct or direct you to vary the work or materials, etc. from that which is contained in the sub-contract documents. Even though the power is there, it is not unfettered. A variation may not be ordered if it changes the whole scope and character of the work. It is easy to think of extreme examples of variation orders which would be invalid under this principle, such as an order omitting an oil-fired boiler driven high pressure hot water system and substituting a warm-air central heating system, but it is difficult to judge less clear examples. The scope of the work may change gradually over the course of several dozen variations, each one of which may seem minor.

To some extent, the matter is academic unless you feel that you are being asked to do something substantially different from what is in your sub-contract and you prefer to stick to the work you priced. Generally, variations mean extra money, extensions of time and reimbursement of loss and/or expense, but if the scope or character of the work is altered, it may result in an unreal or even no proper basis for payment. If you are in this position, you should make your position clear (**document 5.06.1**).

DOM/1 and NSC/C provide for the valuation of variations in alternative clauses: 16 and 17, and 4.4 to 4.9 and 4.10 to 4.13 respectively, depending upon whether the price for the works is to be the sub-contract sum adjusted as necessary or whether the whole of the sub-contract works is to be the subject of remeasurement. NSC/C stipulates that the quantity surveyor will carry out the valuation of variations; DOM/1 does not state who is to do the valuation. Presumably, therefore, it is the contractor. In either case, any queries must be addressed to the other party to your contract, i.e. the contractor. NAM/SC has only one valuation clause: 16.

In general terms, the valuation proceeds on the basis that the priced document provides a starting point to be varied depending on the amount of change from the character, conditions or quantity of the work. In extreme cases, where there is little similarity to priced items, a fair valuation or daywork rates may have to be used. If you consider that the wrong basis of valuation has been used, you must notify the contractor immediately (**document 5.06.2**).

5.06.1   Letter if purported variation changes the scope and character of the work

---

To the Contractor

Dear Sirs,

[Heading]

We are in receipt of your direction number *[insert number]* of the *[insert date]* which you purport to have issued as a variation under the provisions of clause 4.2.1 *[substitute '5.2.1' when using NAM/SC or '3.3.1' when using NSC/C]*.

If we were to comply with your direction, the whole scope and character of the sub-contract works would be changed from those which we contracted to carry out. You have no power to issue such a direction under the terms of the sub-contract and we request that you withdraw it forthwith.

[Add if required:]

If you wish us to carry out this alteration, it must be the subject of a separate contract. We should be delighted to meet you to discuss terms if that is what you require.

Yours faithfully

5.06.2 Letter if the wrong basis of valuation used

To the Contractor

Dear Sirs,

[*Heading*]

We refer to our telephone conversation with your Mr [*insert name*] with regard to the valuation of your direction number [*insert number*] of the [*insert date*].

In our opinion you [*substitute 'the quantity surveyor' when using NSC/C*] have/has [*delete as appropriate*] used the wrong basis of valuation.

[*Briefly describe the reasons why a particular basis of valuation should have been used, e.g. erection of the extra 350 sq m of suspended ceiling in conference area no. 3 will have to take place after the hardwood floor finish has been laid. We contracted, and both your and our programmes show, that all our suspended ceilings were to be erected before the floor finish was put down. We are clearly involved in a more difficult operation and extreme care has to be taken to avoid damage.*]

[*continued*]

5.06.2   *contd*

We should be pleased if proper allowance could be made for these
factors in the rates for the work included in the variation.

Yours faithfully

5.07 Materials, goods and workmanship

DOM/1 and NSC/C have very similar provision in clauses 4.1.2 and 4.1.3, and 1.9.2 and 1.9.3 respectively. Materials and goods shall be, so far as procurable, of the kinds and standards described in the sub-contract documents. Workmanship is to be of the standards described in the sub-contract documents, but if no standards are specified, it is to be of a standard appropriate to the sub-contract works.

These provisions echo the main contract provisions. The words 'so far as procurable' are a valuable protection. If it were not for them, your obligation to provide the goods and materials specified would be absolute. The situation is, however, that if you genuinely cannot obtain the goods or materials for reasons beyond your control, not just because they are more expensive than you expected, you are released from your obligation to provide them. If this situation arises, you should write to the contractor immediately (**document 5.07.1**). You may not use substitute materials without a direction from the contractor. In these circumstances, the direction will be treated as a variation.

There is no such proviso with regard to workmanship. It must be carried out to the standards specified. If no standards are specified, it would be prudent to obtain the contractor's written approval that the standard you have produced is 'appropriate to the sub-contract works'. What is appropriate will be decided, if sweet reason will not prevail, by an arbitrator. **Document 5.07.2** is a suitable letter. If the contractor takes what you consider to be an unreasonable stance, you should record the position even if you then comply with the contractor's demands (**document 5.07.3**).

Both clauses have a proviso which states that if approval of quality of goods and materials or workmanship is a matter for the architect's opinion, they must be to his reasonable satisfaction. There are two important points to remember about this proviso. The first is that if the architect specifies that something is to be to his approval, he must be reasonable about giving his approval. He may not, for example, demand an exceptionally high standard. If you consider that he is being unreasonable, you should write to the contractor and record your views (**document 5.07.4**). It should be noted that there is authority to the effect that the reference is not simply to matters which the architect has expressly reserved for his approval or satisfaction, but that it refers to all matters which are inherently matters for the architect's approval. Such matters will certainly include any matters which are not precisely specified.

The second point, which many architects do not appreciate, is that under clause 21.10.1.1 of DOM/1 and DOM/2 and under clause 19.9 of NAM/SC the final payment, and under clause 4.25.1.1. of NSC/C the final certificate of the main contract, is conclusive that the architect is satisfied with any matters which are to be to the architect's satisfaction. Included will be matters which are inherently matters for the architect's satisfaction as noted earlier. The final payment or final certificate, as appropriate, will be conclusive even if the architect has not even looked at the matters in question. The conclusivity takes effect ten days after the final payment is made in the case of DOM/1, DOM/2 and NAM/SC, and 21 days after the issue of the final certificate in the case of NSC/C. If, therefore, after the expiry of either of these periods you are informed that some of your work is defective, you may be able to refer to clause 21.10.1.1, 19.9 or 4.25.1.1 as the case may be in defence (**document 5.07.5**)

NAM/SC has a very much simpler clause 5.1 which merely states that you must carry out and complete the sub-contract works in accordance with the sub-contract documents and the contractor's reasonable directions. There is no reference to whether materials are procurable; therefore your obligation to supply the materials for which you have quoted is absolute. It is sometimes argued that if such materials are not procurable, the contract would be frustrated, but it is considered that the unobtainable materials would have to be of great significance before frustration would occur. In practice, if they really are not procurable, it probably means that you must obtain the contractor's consent to the use of appropriate substitute materials at no additional cost (**document 5.07.6**). There is a proviso to the clause as before regarding the architect's approval. The final payment is made conclusive about the architect's satisfaction by clause 19.9 as already noted.

One of the employer's main concerns is that when he pays money to the contractor on any certificate, the materials included in that certificate become his property. The lawyers call this 'passing of title' to the employer. It is common knowledge that many suppliers have a retention of title clause in their standard conditions. This usually provides that the supplier does not give up ownership of the materials until he has been paid. The legal implications are quite complex; DOM/1 clause 21.4.5, NSC/C clause 4.15.4 and NAM/SC clause 19.5 are similarly worded in an attempt to ensure proper passing of title to the employer once the contractor has been paid. If you are caught up in a dispute of this nature, you should obtain proper advice without delay.

An important stipulation in each of these clauses is that you are not allowed to remove materials from site once delivered, unless you have the contractor's written consent. Obviously, the contractor would be in a difficult position if you removed permanently materials for which you had received payment, but it may sometimes happen that you have good reason for wanting the materials removed from site. For example, your insurers may demand it. If you have a valid reason for removing materials, use **document 5.07.7** to seek consent.

5.07.1   Letter if goods and materials are not procurable
This letter is not suitable for use with NAM/SC

To the Contractor

Dear Sirs,

[Heading]

We have been informed by our suppliers that *[insert the material]*
specified in the sub-contract is not procurable because   *[insert
reason]*.   We enclose a copy of the letter dated *[insert date]* from
our supplier which explains matters. You will note that our
supplier offers an alternative, albeit at the higher price of *[insert
price]*. A copy of the original quotation we obtained before
tendering is also enclosed.

We should be pleased, therefore, to receive your directions
requiring a variation as a matter of urgency. Your direction must
be in our hands by *[insert date]* at the latest if we are not to be
delayed.

Yours faithfully

5.07.2 Letter if no standards are specified for workmanship
This letter is not suitable for use with NAM/SC

To the Contractor

Dear Sirs,

[*Heading*]

Clause 4.1.3 [*substitute '1.9.3' when using NSC/C*] provides that where no standards for workmanship are specified in the sub-contract documents, the standard must be 'appropriate to the Sub-Contract Works' unless approval of workmanship is a matter for the opinion of the architect. We note that the standard is not specified in the documents.

We have completed the first section of our work in [*describe the location*] and we should be pleased if you would inspect it and confirm that it is of a standard appropriate to the sub-contract works.

Yours faithfully

5.07.3 Letter if the contractor is unreasonable on the standard of workmanship

This letter is not suitable for use with NAM/SC
Registered post/recorded delivery

To the Contractor

Dear Sirs,

[Heading]

We refer to *[describe the work]* in *[describe the location]*. Your site agent has informed us that the standard of workmanship is not satisfactory. In this connection, we have drawn your attention to clause 4.1.3 *[substitute '1.9.3' when using NSC/C]* of the conditions of sub-contract which state that if no standards are specified in the sub-contract documents, the standard must be appropriate to the sub-contract works unless approval of workmanship is a matter for the opinion of the architect.

In our view, the standard of our workmanship is appropriate to the sub-contract works. The workmanship has been inspected by an experienced architect, Mr *[insert name]*, who agrees with our view. We have taken record photographs. Notwithstanding our representations, you have directed us to produce a higher standard of workmanship to precise details as follows:

[continued]

5.07.3   *contd*

[*insert details*]

Take this as notice that we will comply with your direction without prejudice to our rights under this contract and at common law. We expect that you will treat the direction as a variation to be valued under the provisions of clause 16/17 [*delete as appropriate or substitute '4.4/4.10' as appropriate when using NSC/C*]. If you fail to do so we will take appropriate action at that time.

Yours faithfully

5.07.4   Letter if the architect requires too high a standard

To the Contractor

Dear Sirs,

[Heading]

Thank you for your letter of the *[insert date]* from which we understand that the architect is not satisfied with *[describe the work]*. The sub-contract documents specify that this work must be to the approval of the architect. An important part of clause 4.1.3 *[substitute '5.1' when using NAM/SC or '1.9.3' when using NSC/C]* stipulates that the architect's satisfaction must be 'reasonable'. In our view, the architect is being unreasonable in refusing to approve our workmanship and we have engaged an independent consultant architect, Mr *[insert name]*, who shares our view.

We should be pleased, therefore, if you would convey our views to the architect and inform him that if he is not prepared to exercise his duties reasonably as set out in the conditions of sub-contract, we are prepared to take the matter to arbitration.

Yours faithfully

5.07.5   Letter if defects pointed out too late

> *To the Contractor*
>
>
> Dear Sirs,
>
> [*Heading*]
>
> Thank you for your letter of the [*insert date*] informing us that
> [*describe*] is defective. We should draw your attention to the fact
> that the work in question is specifically left for the approval of the
> architect under the provisions of clause 4.1 [*substitute '5.1' when
> using NAM/SC or '1.9' when using NSC/C*].   Clause 21.10
> [*substitute '19.9' when using NAM/SC or '4.25.1' when using
> NSC/C*] stipulates that the final payment [*substitute 'final
> certificate issued under the main contract conditions' when using
> NSC/C*] is conclusive evidence that where the quality of
> workmanship or materials is to be to the architect's satisfaction, it is
> to his satisfaction.
>
> Since evidence now exists that the work is to the architect's
> satisfaction, we suggest any queries be addressed to him and not to
> us.
>
> Yours faithfully

5.07.6   Letter if materials are not procurable
This letter is only suitable for use with NAM/SC

To the Contractor

Dear Sirs,

[*Heading*]

We have been informed by our suppliers that [*insert the materials*]
is no longer procurable because [*insert reason*]. A copy of their
letter dated [*insert date*] is enclosed for your attention.

You will note that they can offer an alternative material which will
give equal service and similar appearance at a cost which is more
than we have included for this item in our sub-contract price. If
you will consent to the use of the alternative material, we are
prepared to supply it at no additional cost to the sub-contract. We
should be pleased to have your consent by [*insert date*] so that we
can place our orders in time to avoid delays.

Yours faithfully

5.07.7   Letter requesting consent to the removal of materials

---

To the Contractor

Dear Sirs,

[*Heading*]

In accordance with the provisions of clause 21. 5.1 [*substitute '19.5.1' when using NAM/SC or '4.15.4.1' when using NSC/C*] of the conditions of sub-contract, we request your written consent to the removal of [*describe the materials*] from site.

The reason for removal is [*insert reasons*].   We intend to relocate the materials at [*insert location*] where they will be clearly identified and open for your inspection.

Yours faithfully

Chapter 6

Certificates and Payment

6.01 Introduction

You are not required to make applications for interim payments, but the contractor will probably expect you to do so and it is certainly in your own interests to submit an application. If you are a nominated sub-contractor under NSC/C, the quantity surveyor will value your work, and the architect, in certifying, will direct the contractor as to the amount included in the certificate in respect of each nominated sub-contractor. Under DOM/1 and DOM/2, clause 21.4.4, you must provide details which are reasonably necessary to substantiate any statement you submit regarding the amount of any valuation. Although this clause requires you to substantiate any application, it does not oblige you to make the application in the first place. If you do not, however, you should not be surprised if you receive rather less than you think is your due.

6.02 Date payable

In the case of DOM/1, DOM/2 and NAM/SC, your first payment is due no later than a month after the date of commencement of sub-contract works on site. Thereafter, payments are due at a maximum of monthly intervals. The contractor is obliged to pay within 17 days of the due date. Sadly, it is common for contractors to ignore these provisions (see section 6.04). Commencement of the sub-contract works on site is a key date because it sets the dates for all future payments. It might appear easy to decide on this date, but in practice, it gives rise to many disputes. The contractor will usually contend that the sub-contract works have not commenced until you are actually on site working. He will usually resist accepting the date of first delivery of your materials to site as the operative date. You should make your position clear (**document 6.02.1**).

Under NSC/C, the position is different. Your payments are linked to the architect's certificates. The contractor must notify you of the amounts included in any architect's certificate as due to you within 17 days of the date of the certificate and he must also pay you within the same period. Quite separately, the architect must also notify you of the appropriate amount due. He has this obligation under clause 35.13.1 of the main contract JCT 80, which, of course, you cannot enforce since you are not a party to its terms. However, under Agreement NSC/W, the architect has the same duty under clause 4 which you can and should enforce (**document 6.02.2**).

The payment terms of NSC/C are very close to being a 'pay when paid' clause. It is certainly a 'pay when certified' clause so that, unless the employer goes into sudden liquidation between certification and payment or for some other reason the main contract is terminated (as in *Scobie & McIntosh Ltd* v. *Clayton Bowmore Ltd* (1990) 49 BLR 119), the likelihood of the contractor being obliged to pay out before he gets paid is remote. If, for any reason, the architect fails to certify you will be in an awkward situation. It is certainly something on which you can seek arbitration under clause 4.20 by using the contractor's name. In that kind of situation, you will find that the contractor is also concerned about the lack of certificate from his own point of view. Arbitration can be protracted and, in a case like this, events can overtake the proceedings. You should, however, send a letter for record purposes (**document 6.02.3**).

6.02.1   Letter if contractor does not agree that delivery of materials is commencement on site

This letter is not suitable for use with NSC/C

To the Contractor

Dear Sirs,

[*Heading*]

We acknowledge receipt of your letter of the [*insert date*] but we do not share your interpretation of 'commencement of the Sub-Contract Works on-site' in clause 21.2.1 [*substitute '19.2.1' when using NAM/SC*].

'Sub-Contract Works' is defined in clause 1.3 as 'the works referred to in the Appendix, part 2, and described in the Numbered Documents ...' [*substitute 'the works referred to in NAM/T, Section 1, and described in the Numbered Documents' when using NAM/SC*]. The materials we delivered to site on [*insert date*] were certainly part of the works so referred to and described.   If you contend that the materials are not such a part, you are saying that we have no obligation under the contract to supply them.

[*continued*]

6.02.1 *contd*

It is our view that you are confusing 'Sub-Contract Works' with 'work' as in 'work and materials'. We should be pleased to hear by return that you accept that, for the purposes of clause 21.2.1 [*substitute '19.2.1' when using NAM/SC*], delivery of materials to site is commencement of the Sub-Contract Works.

Yours faithfully

6.02.2 Letter if the architect does not notify you of sub-contract amounts included in a certificate

This letter is only suitable for use with NSC/C

To the Employer
(Copy to the Architect)

Dear Sirs,

[Heading]

We refer to the JCT Standard Form of Employer/Nominated Sub-Contractor Agreement NSC/W which we entered into with you on *[insert date]*.

By clause 4 you undertake that the architect will operate the provisions of clause 35.13.1 of the main contract. Part of clause 35.13.1 states that the architect will forthwith inform each nominated sub-contractor of the amount of any interim or final payment about which he has directed the contractor in accordance with clause 35.13.1.1. The architect did not so inform us in regard to the amount included in the last certificate and, therefore, we cannot check whether the contractor has made the proper payment to us.

We are assuming that in this instance there has been an oversight and we should be pleased if you would arrange to correct the matter immediately. If the situation is not corrected within two days from the date of this letter or if the situation occurs again, we shall consider the architect's failure to inform us as a serious breach of contract about which we will take appropriate advice.

Yours faithfully

6.02.3   Letter if architect fails to certify

This letter is only suitable for use with NSC/C

Registered post/recorded delivery

To the Contractor

Dear Sirs,

[Heading]

We refer to our telephone conversation this morning when you informed us that you have not paid us because the architect has not yet issued a certificate this month.

The architect's reasons for withholding his certificate are irrelevant. Under clause 4.16.1.1 of the conditions of sub-contract our payment is clearly linked to the issue of interim certificates under the main contract. We are now incurring financing charges which we will seek to recover. Take this as notice, therefore, that unless we receive payment within seven days from the date of this letter, we may require you to allow us to use your name and join with us in arbitration proceedings in accordance with the provisions of clause 4.20.

Yours faithfully

6.03   Amount payable

The amount payable is to be calculated under DOM/1 and DOM/2 by applying the provisions of clause 21.4 to arrive at the amount properly due, then deducting certain amounts specified in clause 21.3. Two amounts may be deducted: the cash discount and the retention. Although the cash discount is only to be deducted if payment is made within 17 days of the date it becomes due, contractors often make the deduction no matter how late the payment is made. You must respond to this treatment (**document 6.03.1**). Another frequent miscalculation is when the contractor deducts the cash discount before deducting the retention. By the time the last of your retention has been released, the figures should have corrected themselves, but until then the contractor has had the use of a greater proportion of your money than was intended. This is another reason for a strong letter (**document 6.03.2**). The wording of clause 21.3 is partly responsible for the muddle. The equivalent clause 19.3 of NAM/SC is much clearer. A slightly different letter is indicated (**document 6.03.3**). The situation should not arise under NSC/C, because the retention is held by the employer before the amount due is included in a certificate.

6.03.1   Letter if cash discount is deducted although payment made late
This letter is not suitable for use with NSC/C

To the Contractor

Dear Sirs,

[Heading]

We note that from your last payment which you made on the *[insert date]* you deducted the sum of *[insert amount]* which you described as 'cash discount'.

We draw your attention to clause 21.3.2 *[substitute '19.3.2' when using NAM/SC]* which states that cash discount can be deducted only 'if payment is made as provided in clause 21.2' *[substitute '19.2' when using NAM/SC]* which provides, among other things, that payments must be made not later than 17 days after they become due.

You can do no other than accept that payment is due on the *[insert date]* of each month since it is common ground that we commenced the Sub-Contract Works on site on the *[insert date]*.

Referring to our first paragraph, it is clear that you made payment *[insert number]* days after payment was due. Cash discount is, therefore, not deductible and we require your cheque to be in our hands by *[insert date]* or we shall take appropriate action.

Yours faithfully

6.03.2   Letter if the contractor deducts cash discount before retention
This letter is only suitable for use with DOM/1 and DOM/2

To the Contractor

Dear Sirs,

[*Heading*]

We refer to the payment of [*insert amount*] which you made to us
on the [*insert date*]. We also refer to our telephone conversation
with your Mr [*insert name*] today when we were informed that, in
your view, cash discount had been correctly deducted before
retention had been taken off.

The contractual position is as follows: Clause 21.3 stipulates that
an interim payment must be the gross amount in clause 21.4 less
only retention, cash discount and the amount previously paid.
According to clause 21.3.2, cash discount may be deducted 'if
payment is made as provided in clause 21.2'. Cash discount is just
what it says: discount for cash. Payment cannot be said to have
been made on that part of the payment which has been retained
under the terms of clause 21.3.1. Payment of the money retained
will not be made and cash discount will not be deductible until the
retention is released.

[continued]

6.03.2 *contd*

Although the final sum payable will eventually be correct under the system you are operating, the effect is to deny us a substantial sum of money until much later than the contract stipulates. This is a clear breach of contract for which you are liable in damages. Some at least of those damages will be the financing charges which we are having to find in respect of the discount wrongly deducted.

We should be pleased to receive the sum of *[insert amount]*, being the money unlawfully withheld, within seven days of the date of this letter or we will look to our legal rights and remedies.

Yours faithfully

6.03.3   Letter if the contractor deducts cash discount before retention
This letter is only suitable for use with NAM/SC

To the Contractor

Dear Sirs,

[Heading]

We refer to the payment of *[insert amount]* which you made to us on the *[insert date]*. We also refer to the telephone conversation with your Mr *[insert name]*, when we were informed that, in your view, cash discount had been correctly deducted before retention was taken off.

The contractual position is as follows: Clause 19.3 stipulates that each interim payment must be the amount calculated in accordance with clause 19.4 less the amount previously calculated for the last interim payment and the cash discount. To arrive at the amount calculated in accordance with clause 19.4, retention has to be taken into account. Thus retention will have been deducted before the deduction of cash discount under clause 19.3.2. Moreover, cash discount may be deducted only if 'payment is made as provided in clause 19.2'. Payment cannot be said to have been made on that part of the payment which has been retained. Payment will not be made on that part until retention is released at the appropriate time.

[continued]

6.03.3 *contd*

Although the final sum payable will eventually be correct, the effect of your system is to deny us the use of money rightfully ours. This is a clear breach of contract for which we are entitled to damages. Part of these damages will be the financing charges we have to find in respect of the money wrongfully deducted.

Unless we receive the sum of [*insert amount*], being the sum unlawfully withheld, within seven days of the date of this letter we will take appropriate legal action to recover the sum forthwith.

Yours faithfully

6.04   Failure to pay

What if the contractor fails to pay or fails to pay all the amount which you consider or, in the case of NSC/C, you have been notified is due? Under DOM/1, DOM/2 and NAM/SC, clauses 21.6 and 19.6.1 respectively, the position is very similar. You have a right under the sub-contract which the contractor does not have under the main contract. You have the right to suspend work. You must first be sure that the contractor has indeed failed to pay as provided by the terms of the sub-contract.

Such failure would include wrongful withholding of cash discount or simply failure to pay at all unless the money withheld was withheld in accordance with the terms of the sub-contract, for example, in respect of legitimate set-off. It is obviously easier to show that the contractor has failed to make payment if he has made no payment at all. If he has made some payment, he can always say that the amount paid is the amount he has valued as due. Unless the underpayment is gross, it is very difficult to suspend work as a result. You should, however, put the facts on record and threaten suspension if you have the evidence to back up your contentions (**document 6.04.1**).

Commonly, the contractor will fail to pay at the right time in accordance with the provisions of clause 21.2 of DOM/1 and DOM/2 or clause 19.2 of NAM/SC. You could try a letter before giving formal notice of suspension (**document 6.04.2**). If that produces no satisfactory response, a formal notice is indicated (document 6.04.3). If the contractor does not pay within seven days as stipulated, you may then suspend work until payment is made. If you do decide to take this major step it is advisable, although not strictly necessary, to confirm your action in writing (**document 6.04.4**). Suspension of work is the best pressure you can put on a contractor, but you must be sure that your ground is certain. If not, it is doubtful that the contractor could treat your action as repudiation, because suspension implies an intention to continue some time in the future, although in some cases it might amount to repudiation and expose you to a claim for damages.

Under NSC/C, you have the right to suspend, but only if the principal remedy does not produce payment. The principal remedy is your right to receive direct payment from the employer if the contractor does not discharge the amount shown payable to you in any certificate. Every time you receive payment from the contractor, clause 4.16.1.1 stipulates that you must furnish him with reasonable proof that he has discharged the payment in accordance with the terms of NSC/C. Generally, this will take the form of a receipt which the contractor will ask you to sign and date. The contractor must give this proof to the architect before the issue of each interim and the final certificate. If you do not agree that the contractor is properly discharging his obligations with regard to payment, you can simply not return the receipt, but you would be prudent to confirm your disagreement in writing with a copy to the architect and the employer (**document 6.04.5**).

Where you have entered into agreement NSC/W with the employer, the architect must operate clauses 35.5.3 and 35.5.4 of the main contract. These stipulate that, if reasonable proof is not provided by the contractor, the architect must certify accordingly to the employer with a copy to you. The employer is obliged to pay you directly and reduce the contractor's certificate by an appropriate amount. Two situations could arise, which you should deal with by letter: if the architect does not issue a clause 35.13.5.2 certificate to the employer (**document 6.04.6**); if the employer does not pay you (**document 6.04.7**).

If, despite these provisions and your letters, you still do not receive payment or the whole of the payment to which you are due, you are entitled to suspend work after giving 14 days' notice to the contractor until you are paid by either the contractor or the employer. **Documents 6.04.3 and 6.04.4** are applicable with the alterations noted.

6.04.1   Letter if contractor does not pay the amount due
This letter is not suitable for use with NSC/C

To the Contractor

Dear Sirs,

[Heading]

Under the terms of our sub-contract, the total amount due to us for each interim payment is to be calculated in accordance with clause 21.3 *[substitute '19.3' when using NAM/SC]*. Our application for payment dated *[insert date]* was strictly so calculated in the sum of *[insert amount]*. Your payment received today was *[insert amount]* short of the proper amount. You have given us no explanation, neither have you given any proper breakdown from which we could identify the cause of the discrepancy. When we telephoned your office today, Mr *[insert name]* declined to give us any further information.

In these circumstances, we have the right to take action under clause 21.6 *[substitute '19.6.1' when using NAM/SC]* and we have our common law rights and remedies. We look forward to receiving your cheque for *[insert amount]* by return.

Yours faithfully

6.04.2 Letter if contractor fails to pay in accordance with the payment terms

This letter is not suitable for use with NSC/C

To the Contractor

Dear Sirs,

[*Heading*]

[*Either:*]

We have today received your payment notification dated [*insert date*]. It is significantly less than the sum to which we are due under the terms of the sub-contract.

[*Or:*]

It is now two days since the date on which we should have received an interim payment.

[*Then:*]

If the situation is not rectified within four days of the date of this letter, we intend to serve formal notice under the provisions of clause 21.6 [*substitute '19.6.1' when using NAM/SC*].

Yours faithfully

6.04.3   Letter if contractor fails to make payment
Registered post/recorded delivery

To the Contractor
(Copy to Architect and, under NSC/C, to the Employer)

Dear Sirs,

[Heading]

Interim payment was due to us on the *[insert date]*. You failed to make such payment. We drew your failure to your attention on the *[insert date]* but you have not made payment at the date of this letter.

Take this as formal notice given under the provisions of clause 21.6 *[substitute '19.6.1' when using NAM/SC or '4.21.1' when using NSC/C]* that if, within seven *[substitute '14' when using NSC/C]* days of the giving of this notice, you do not discharge your obligation to make payment to us as provided in the sub-contract, it is our intention to suspend the further execution of the Sub-Contract Works until such discharge occurs.

Yours faithfully

6.04.4 Letter suspending work
Registered post/recorded delivery

To the Contractor
(Copy to Architect and, under NSC/C, the Employer)

Dear Sirs,

[*Heading*]

We refer to our notice dated [*insert date*] given under the
provisions of clause 21.6 [*substitute '19.6.1' when using NAM/SC
or '4.21.1' when using NSC/C*]. Seven [*substitute '14' when using
NSC/C*] days have passed since the giving of such notice and you
have not discharged your obligation to make payment as specified
in such notice.

We, therefore, exercise our right from today to suspend further
execution of the sub-contract works until you have discharged your
obligation to make payment in accordance with the terms of the
sub-contract [*add: 'or direct payment has been made by the
employer' when using NSC/C*].

Such suspension is not to be deemed a failure on our part to
proceed with the Sub-Contract Works in accordance with the
provisions of the sub-contract. This action is without prejudice to
any other rights and remedies which we may possess.

Yours faithfully

6.04.5   Letter if contractor does not pay
This letter is only suitable for use with NSC/C

To the Contractor
(Copies to Architect and Employer)

Dear Sirs,

[*Heading*]

We refer to the cheque and payment notification which we received
from you on [*insert date*]. The sum shown is [*insert amount*]
short of the amount notified to us by the architect under the
provisions of clause 35.13.1.2 on the [*insert date*] as interim
payment to us included in the current certificate.

You have seven days from the date of this letter to pay us the
shortfall noted above, failing which we will require the employer
under clause 7.1 of the JCT Standard Form of
Employer/Nominated Sub-Contractor Agreement to exercise his
right under the main contract clause 35.13 to make direct payment
to us.

Yours faithfully

6.04.6 Letter to the employer if the architect does not issue a main contract clause 35.13.5.2 certificate

This letter is only suitable for use with NSC/C

To the Employer
(Copy to Architect)

Dear Sirs,

[Heading]

We refer to our letter of the *[insert date]* addressed to the contractor, in which we recorded the fact that the contractor had failed to pay us *[insert amount]* due from the last certificate.

The architect has not operated clause 35.13 of the main contract to the extent that he has not yet issued a clause 35.13.5.2 certificate which is a pre-condition to you making direct payment to us. We, therefore, require you to fulfil your obligations under clause 7.1 of the JCT Standard Form of Employer/Nominated Sub-Contract and together with the architect operate the provisions of clause 35.13.

Yours faithfully

6.04.7   Letter to the employer if he does not pay direct
This letter is only suitable for use with NSC/C

To the Employer
(Copy to Architect)

Dear Sirs,

[*Heading*]

We have received a copy of a certificate dated [*insert date*] issued
by the architect under clause 35.13.5.2 of the main contract. Under
the terms of the JCT Standard Form of Employer/Nominated
Sub-Contractor, clause 7.1, you are now required to operate the
provisions of clause 35.13 of the main contract and make direct
payment to us of the amount the contractor has failed to discharge.

It is now ten days since the date of the architect's clause 35.13.5.2
certificate and we regret that if you do not make payment as
indicated above within seven days from the date of this letter we
will take appropriate action.

Yours faithfully

6.05   Set-off

Set-off is the single biggest source of dispute between contractors and sub-contractors. This is generally because the sub-contractor contends that the contractor is deducting money for insufficient reason or for no reason at all. All three sub-contract forms provide for the contractor to set off sums against amounts due to the sub-contractor under the payment terms. In all cases, the set-off rights are expressly stated to be fully set out in the sub-contract and no other rights of set-off are to be implied. However a law case in 1989 (*Acsim (Southern) Ltd* v. *Dancon Ltd* (1989) 19 Con LR 1) established that the set-off clauses do not contain the exclusive machinery by which the contractor can challenge your right to interim payment on grounds that you have not properly executed the work or his right to raise other legal defences.

The set-off provisions are contained in clause 23 (DOM/1 and DOM/2), clause 4.26 to 4.29 (NSC/C) and clause 21 (NAM/SC). They are very similar. The contractor is entitled to deduct from any money otherwise due amounts agreed by the sub-contractor or awarded in arbitration or litigation under the sub-contract. That is straightforward and rarely gives rise to disputes. Difficulties arise over the second part of the set-off clauses. This provides that the contractor is entitled to set-off any amount of loss and/or expense and/or damage suffered or incurred by the contractor as a result of the sub-contractor's breach of the sub-contract. There are important stipulations which cause problems for many contractors whilst providing valuable protection for the sub-contractor:

- The amount of the set-off must be quantified in detail and with reasonable accuracy by the contractor; and
- The contractor must give the sub-contractor written notice which specifies his intention to set off the amount so quantified, together with the details and the grounds upon which the set-off is made.

Under DOM/1, DOM/2 and NAM/SC, the notice must be given not later than three days before the payment, from which the contractor intends to set off, becomes due. The payments are made at monthly intervals from the date of commencement on site; therefore, the contractor should realise that he must issue the notice at least three days before the monthly date. Since payment must be made no later than 17 days after the due date, the notice gives you 20 days in all to take action. Since the adjudication provisions follow a strict timetable, it is important that the contractor adheres to the notice provisions. Failure to do so will disentitle the contractor to set off the amount from the next payment. If the contractor gives late notice, you must respond immediately (**document 6.05.1**).

If the contractor fails to quantify with sufficient detail or accuracy, you must dispute the notice, although it may be given on time (**document 6.05.2**). It is not always easy to decide whether sufficient detail has been given or whether the calculation has been carried out with reasonable accuracy. In the last analysis, it is a matter for the judge, arbitrator or adjudicator.

There are, however, some simple guidelines which you can apply. The contractor must not simply put down one figure, say, £10 000. He must demonstrate how he has arrived at the figure. He must go further and show where he has obtained the figures he uses to build up his total. The calculation need not be accurate to the nearest pound, but the contractor may not simply take a stab at what he imagines he may have lost.

Importantly, the clause refers to loss and/or expense and/or damage suffered or incurred. This precludes the contractor from setting off anticipated losses. He may well be 99% sure that the employer will deduct liquidated damages which he will want to pass on to you. It may well be the case that you are culpable, but until the contractor suffers or incurs the loss rather than the liability for the loss, he may not use this clause to deduct it from you. It should be noted that the right of the contractor or sub-contractor to vary the grounds and amount of set-off in any subsequent proceedings is expressed reserved.

Under NSC/C, notice must be given at least three days before the date of issue of the interim certificate which includes an amount in favour of the sub-contractor from which the contractor intends to make the set-off. This again gives you 20 days in which to take action. The provision, in fact, puts the contractor at something of a disadvantage because he can never be absolutely sure of the date of issue of any interim certificate three days in advance of its issue. Nor can he be sure of the amount to be included in favour of any nominated sub-contractor.

To be safe, if the contractor wishes to set off against you under this sub-contract, he must give notice several days before it is due. If he miscalculates and the interim certificate is issued only two days later, you have solid grounds to dispute the notice (**document 6.05.3**). No loss and/or expense relating to delay in completion can be set off unless there is in existence a non-completion certificate issued by the architect in accordance with clause 35.15 of the main contract and clause 2.9 of NSC/C.

6.05.1 Letter if contractor gives late notice of set-off

This letter is not suitable for use with NSC/C

To the Contractor

Dear Sirs,

[Heading]

We are in receipt of your letter dated *[insert date]* which purports to be a notice of set-off under the provisions of clause 23.2 *[substitute '21.2' when using NAM/SC]*.

Any such notice is to be given not less than three days before the date upon which the payment from which you intend to make the set-off becomes due.

Payment is due on the *[insert date]*, therefore your notice is invalid and should be withdrawn. If you attempt to set-off the sum indicated in your purported notice, we shall take immediate legal proceedings for recovery.

Yours faithfully

6.05.2   Letter if notice of set-off not quantified in detail
This letter is not suitable for use with NSC/C

To the Contractor

Dear Sirs,

[*Heading*]

We are in receipt of your letter of the [*insert date*] which purports
to be a notice of set-off under the provisions of clause 23.2
[*substitute '21.2' when using NAM/SC*].

In any such notice the set-off must be quantified in detail and with
reasonable accuracy. Your purported notice makes no real attempt
to satisfy this requirement. Therefore, it is invalid and should be
withdrawn. If you attempt to set off the amount you have
indicated, we shall take immediate legal action for recovery.

Yours faithfully

6.05.3 Letter if the contractor gives late notice of set-off
This letter is only suitable for use with NSC/C

To the Contractor
(Copies to Architect and Employer)

Dear Sirs,

[*Heading*]

We have received your letter dated [*insert date*] which purports to be a notice of set-off under the provisions of clause 4.27.

Any such notice is to be given not less than three days before the date of issue of the interim certificate which includes, in the amount stated as due, an amount in our favour and from which amount you intend to make the set-off.

The interim certificate to which your notice refers was issued on the [*insert date*]. As a matter of fact your purported notice was issued late and it is, therefore, invalid and should be withdrawn. If you attempt to set off the amount you have indicated, we shall take immediate steps for recovery.

Yours faithfully

6.06   Adjudication

The adjudication provisions are to be found in clause 24 of DOM/1 and DOM/2, clause 22 of NAM/SC and clause 4.30 to 4.37 of NSC/C. They are in very similar terms. If the contractor sends you notice that he intends to set-off a certain amount against your next payment and you disagree with either the whole or any part of the amount, you must send your reasons for disagreement to the contractor in the form of a written statement, together with particulars of any counterclaim, which you, in turn, must quantify in detail and with reasonable accuracy (**document 6.06.1**). You may amend the document later if the disagreement is to be decided by arbitration. At the same time as you send the statement, you must give the contractor notice of arbitration (see Chapter 11) and you must request action by the adjudicator (**document 6.06.2**), notifying the contractor of your request.

The name of the adjudicator is to be inserted in the contract documents. If the space is merely left blank, the clause provides that the adjudicator is to be a person appointed by you from the list of adjudicators maintained by the Building Employers Confederation. It is common (but bad) practice, to fill in the space with the words 'to be agreed'. On a strict reading of the contracts, it appears that, under NSC/C, the conditions will prevail and the effect will be the same as if the space was left blank (i.e. the adjudicator is not named). Under DOM/1, DOM/2 and NAM/SC, the position is that the words 'to be agreed' prevail and, if you receive a notice of set-off, it is thought that the notice may be invalid unless an adjudicator can be agreed in time to permit you to send the appropriate request to him within the 14 days allowed by the contract. Since you will be one of the agreeing parties, it is not likely, to say the least, that you will rush to take an action which will validate a set-off notice against you. If you do receive a notice under these circumstances, you should reply immediately (**document 6.06.3**).

The contractor has 14 days from receipt of your statement to respond to the adjudicator. The adjudicator has a further seven days in which to decide whether the whole or part of the disputed amount:

- is to be retained by the contractor; or
- must be deposited with a trustee-stakeholder pending arbitration; or
- must be paid to you; or
- any combination of the above.

There the matter rests until and unless the arbitrator decides otherwise. The arbitrator is permitted to cancel or vary the adjudicator's decision at any time before his final award, if either party makes an application to him.

If the contractor refuses to honour the adjudicator's decision, you should seek legal advice with a view to obtaining an injunction to compel him to do so.

6.06.1   Letter if sub-contractor disagrees with set-off
Registered post/recorded delivery

To the Contractor

Dear Sirs,

[*Heading*]

We are in receipt of your written notice dated [*insert date*], issued
under clause 23.2 [*substitute '21.2' when using NAM/SC or '4.27'
when using NSC/C*] of the conditions of sub-contract, in which you
state your intention to set-off the sum of [*insert amount*] from our
next payment.

In accordance with the provisions of clause 24 [*substitute '22' when
using NAM/SC or '4.30' when using NSC/C*] we enclose our
written statement setting out our reasons for disagreeing with the
amount specified in your notice.   [*Add 'We also enclose particulars
of our counterclaim against you which we have quantified in detail
and with reasonable accuracy', if appropriate*].

We have today requested action by the adjudicator.

Enclosed is a notice of arbitration as required by clause 24.1.1.1
[*substitute '22.1.1.1' when using NAM/SC or '4.30.1.1' when using
NSC/C*].

Yours faithfully

6.06.2   Letter requesting action from the adjudicator
Registered post/recorded delivery

To the Adjudicator

Dear Sir,

[Heading]

You are named in the Appendix, part 8 [*substitute 'Article 3.1'
when using NAM/SC or 'NSC/T, Part 3, item 3' when using
NSC/C*] to this sub-contract as adjudicator under the provisions of
clause 24 [*substitute '22' when using NAM/SC or 'clauses 4.30 to
4.37' when using NSC/C*]. We hereby request you to act in
accordance with the right given in clause 24.1.2 [*substitute '22.1.2'
when using NAM/SC or '4.30.2' when using NSC/C*].

For this purpose we enclose our written statement dated [*insert
date*] and the contractor's notice of the [*insert date*] to which our
statement relates [*add 'and our counterclaim' if appropriate*],
copies of which have been sent to the contractor.

Yours faithfully

6.06.3 Letter if set-off notice received and no adjudicator appointed
This letter is not suitable for use with NSC/C
Registered post/recorded delivery

To the Contractor

Dear Sirs,

[Heading]

We have received your letter dated *[insert date]* which purports to
be a notice of set-off. In these circumstances, it would be our
intention to request the adjudicator to act under the provisions of
clause 24 *[substitute '22' when using NAM/SC]*.

In this sub-contract no adjudicator has been named in the
Appendix, part 8 *[substitute 'Tender, Schedule 2, item 6' when
using NSC/4 or 'Article 3.1' when using NAM/SC]*. Instead, the
words 'to be agreed' have been inserted. The contract must be
read as a whole and particularly clauses 23 and 24 *[substitute '21
and 22' when using NAM/SC]* must be read together. Your right
to set-off is clearly dependent upon our right to seek adjudication
on the matter within the 14 days allowed.

We, therefore, request you to put forward the name of a proposed
adjudicator. It is our view that failure to reach agreement in time
to allow us to operate the provisions of clause 24 *[substitute '22'
when using NAM/SC]* will render your notice of set-off invalid.

Yours faithfully

6.07 Final payment provisions

Under DOM/1 and DOM/2, you must send the contractor all documents necessary for the adjustment of the sub-contract sum or the computation of the ascertained final sub-contract sum as applicable, depending upon whether clause 15.1 or clause 15.2 (complete remeasurement) apply. You must send the information not later than four months after practical completion of the sub-contract works. The rules governing the final calculations are contained in clauses 21.7 and 21.8. The amount of the final payment is often a source of dispute.

Although there is no express term in the contract, it is common practice for the contractor to write to the sub-contractor requesting agreement to the final sub-contract account. If you receive such a letter and you do not agree with the sum, do not simply disagree; you should submit your alternative calculation (**document 6.07.1**). The contractor will sometimes go too far, stipulating a sum of money which he admits is due and requesting you to agree to accept it in full and final settlement of his obligations under the sub-contract. It may be that the sum is many thousands of pounds less than you anticipated. It is worthwhile trying **document 6.07.2**.

The final payment is due not later than seven days after the date of the final certificate issued by the architect under the provisions of the main contract JCT 80. Before that date, the contractor must send you a notice to inform you of the amount of the payment you will receive. If you do not agree with the amount, you must again write and put your position forcibly. **Document 6.07.1** is applicable with minor amendments. The contractor must make payment within 28 days of the due date. It should be noted that the final payment is conclusive evidence that:

- Where the quality of materials or goods and the standard of workmanship are matters for the architect's satisfaction, the materials, goods and workmanship in question are to his reasonable satisfaction.
- All sub-contract clauses governing the adjustment of the sub-contract sum have been complied with except for arithmetical error or the accidental inclusion or exclusion of work, materials or figures.
- All extensions of time which are due have been given.
- Final reimbursement of loss and/or expense has been made in settlement of all claims arising from matters under clause 13.1.2 in whatever form they are presented.

The only exceptions to the conclusiveness of the final payment is if you have started proceedings before a date ten days after the contractor's notice or ten days after the final payment, whichever occurs first. There is, of course, little doubt that the notice will be the first to be received and it is, therefore, important to remember that you must not wait until you actually receive payment. If you dispute the final payment, you must serve notice of arbitration immediately (see Chapter 11). You must not simply write and complain. At this stage nothing short of a notice of arbitration or, of course, a writ will suffice.

Under NAM/SC, the position is similar to DOM/1. You must send documents to the contractor within five months of the date of practical completion of the sub-contract works and the final payment is conclusive about the same matters unless proceedings have been commenced within ten days after the date of the contractor's notice or the date of the final payment, whichever is latest. **Documents 6.07.1 and 6.07.2** are applicable.

If you are a nominated sub-contractor having entered into NSC/C, a strict timetable applies similar to that leading up to the issue of the final certificate under the main contract. You must send all the necessary documents for adjustment of the sub-contract sum or for the purpose of computing the ascertained final sub-contract sum to the contractor no later than six months after practical completion of the sub-contract works. Alternatively, the contractor may instruct you to send the documents to the architect or the quantity surveyor under the main contract. The rules for calculating the final sub-contract sum are contained in clauses 4.23 and 4.24, depending upon whether NSC/A article 3.1 or 3.2 applies.

No later than three months after receipt of all the necessary documents, the architect or the quantity surveyor must send you a statement showing how the final sub-contract sum has been calculated. He must do this before your final payment is certified. There is no specific provision for disagreement, but if you disagree, **document 6.07.1** is applicable.

The actual final payment of a nominated sub-contractor is governed by clauses 30.7 or 35.17 of the main contract. The basic rule is that the architect must issue an interim certificate at least 28 days before he issues the final certificate under the main contract. The interim certificate must contain the finally adjusted or ascertained sub-contract sums of all the nominated sub-contractors. If, however, you have entered into the form of Employer/Nominated Sub-Contractor Agreement NSC/W, your position is better. The architect may issue the interim valuation at any time after the date of practical completion of the sub-contract works and he must issue it within 12 months. There are two important provisos:

- The architect and the contractor must agree that you have remedied any defects which have appeared and which you are obliged to remedy under the terms of the sub-contract.
- You have sent all the necessary documents for the final calculations of the sub-contract sum to be made.

If the architect does not issue the certificate after the expiry of the 12 month period, or before that if you have rectified all the defects in your work, you should write to the employer (**document 6.07.3**). The response will probably come from the architect. If the response is that you have not rectified all the defects in your sub-contract work, you should make your position clear (**document 6.07.4**). If neither the architect nor the contractor will move, you must consider arbitration if the amount of money at stake is sufficiently large.

Unusually, the final certificate issued under the main contract is conclusive evidence under the sub-contract. It is conclusive about the same four matters as is the final payment under DOM/1. The exception is if arbitration or other proceedings have been started no later than 21 days after the issue of the final certificate.

6.07.1　Letter if the sub-contractor does not agree with the final sub-contract account

Registered post/recorded delivery

To the Contractor
(Copy to the Architect and the Employer when using NSC/C)

Dear Sirs,

[*Heading*]

We have received notice dated [*insert date*] of the total amount of the final payment/final account/ascertained final sub-contract sum [*delete as appropriate*].

We do not agree with the total amount shown as payable to us and we attach our own calculation of what we believe is the true figure. We are hopeful that you will agree with us and we are, naturally, prepared to discuss the matter with you.

In order to protect our interests, we will be obliged to serve notice of arbitration if agreement has not been reached within five days of the date of the above notice.

Yours faithfully

6.07.2 Letter if contractor offers a sum in full and final settlement

To the Contractor

Dear Sirs,

[Heading]

Thank you for your letter of the *[insert date]* from which we note that you acknowledge that you owe us the sum of *[insert amount]* for the work we have carried out under this sub-contract.

We are not prepared to accept such sum in full and final settlement as you suggest, because we are entitled to considerably more than you offer. Nevertheless, we should be pleased to receive the sum you note, as partial payment, within seven days from the date of this letter.

If we do not receive payment of the sum which, by the evidence of your own letter, is indisputably due, we shall apply to the Court for summary judgment in the matter.

Yours faithfully

6.07.3   Letter if architect does not issue an interim certificate including the finally adjusted or ascertained sub-contract sum

This letter is only sustainable for use with NSC/C

To the Employer
(Copy to Architect)

Dear Sirs,

[*Heading*]

We refer to the JCT Standard Form of Employer/Nominated Sub-Contractor Agreement, clause 5.1 of which stipulates that the architect will operate clauses 35.17 to 35.19 of the main contract conditions.

It is now 12 months since the date of practical completion of our sub-contract works and although the architect should by now have issued an interim certificate the gross valuation of which must include the amount of the sub-contract sum as finally adjusted [*substitute 'as finally ascertained' if appropriate*], he has not yet done so.

This is nothing short of breach of contract which we now call upon you to remedy within seven days from the date of this letter. Should you fail to take action, we will pursue our other remedies.

Yours faithfully

6.07.4   Letter if contractor contends that defects are not rectified
This letter is only suitable for use with NSC/C

To the Architect
(Copy to Employer)

Dear Sirs,

[Heading]

We are in receipt of your letter of the *[insert date]* in response to
our letter of the *[insert date]* addressed to the employer.

We do not accept that you are entitled to withhold certification of
the final payment to us on the grounds that we have not completed
all the rectification of defects in our sub-contract work.

We have rectified all defects, shrinkages and other faults which
have appeared and which we are bound to remedy under our
sub-contract. You have not demonstrated to the contrary. If you
do not properly operate the provisions of clause 35.17 of the main
contract and include our final payment in your next interim
certificate, there will be a breach of contract for which the
employer will be liable. We are incurring losses and financing
charges and we will seek to recover appropriate damages.

Yours faithfully

Chapter 7

Delays and Extensions of Time

7.01   General principles

Under the general law, your obligation is to complete the sub-contract works by the date agreed in the sub-contract. This obligation will be removed if you are prevented or hindered from completing by the fault of the contractor or the employer. In those circumstances, you would simply have to complete within a reasonable time: this is commonly referred to as time being 'at large'. Since it is almost inevitable that the contractor will do something which will prevent you from completing on the due date, such as ordering extra work or providing information late and so on, the result would be that every contract would lose its fixed completion date. This would make it difficult, although not impossible, for the contractor to recover any damages from you for late completion. Naturally, that would be something very much in your favour.

The contractual answer to the problem is to make provision for the sub-contract period to be extended on account of the contractor's or the employer's faults. In this way there is always a completion date which is fair and reasonable. All the extension of time clauses also allow the sub-contract period to be extended on account of certain events which are certainly not the fault of either contractor or employer, but which are not your fault either. Such things as exceptionally adverse weather conditions are included in this category.

7.02   Possession

Clause 11 deals with extensions of time in DOM/1 and DOM/2. Clause 12 deals with it in NAM/SC and clauses 2.2 to 2.7 in NSC/C. The clauses are similar, but not identical. Your obligation to carry out and complete the sub-contract works by the date for completion is spelled out, so is your obligation to commence on the date stated in the sub-contract (see section 5.01). In each case, however, commencement is made subject to the receipt of notice to commence work as detailed in the sub-contract and it is this notice which determines the contractual date for commencement. Therefore, if two weeks notice is specified, you cannot be made to begin work on site before the two weeks has expired. The date for commencement of work, whether in the sub-contract or in a notice to begin, is the date for possession. It is the date on which you are entitled to have possession of the site or at least that part of the site which is necessary to enable you to carry out your work.

Under the provisions of the main contract, failure to give possession on the due date is a serious breach of contract unless the deferment provisions have been operated. In general, they allow possession of the site to be deferred for up to six weeks. The facility is useful if the employer is prevented from giving possession for some reason. There is no equivalent provision 'stepped down' to the sub-contracts and the notice provisions should make it unnecessary, but there is provision for extension of the sub-contract period if the contractor's possession of the site is deferred. If, however, you receive notice to commence on a certain date, that is the date when you should have possession and if, for any reason, possession is not given, you would have grounds for extension of time on the basis of 'act, omission or default' of the contractor. You would also have grounds for claiming loss and/or expense.

7.03 Procedures

You have certain duties which must be carried out whenever your sub-contract works are delayed for any reason. You must give notice in writing to the contractor whenever it becomes reasonably apparent that the commencement, progress or completion of the sub-contract works is likely to be delayed. You must include in your notice the cause of the delay insofar as you are able and you must identify anything which you think entitles you to an extension of time under the contract. Note that you must notify all delays and not just delays which you consider entitle you to an extension of time. **Document 7.03.1** is a letter notifying the contractor if the delay does not entitle you to an extension of time. **Document 7.03.2** is the kind of letter you should write if you think the delay does entitle you to an extension of time.

Your duties do not end at that point. If you have identified to the contractor an event which you consider entitles you to an extension of time you must, as soon as possible, send further particulars (**document 7.03.3**). These particulars include the effects of the event and an estimate of the delay in completion of the sub-contract works. For example, the delay may be caused by the contractor's direction that extra work is required. It may well be crystal clear that delay will result, but you may not be able to foresee the consequences or to estimate the extent until after you have completed the work. Do not fall into the trap of estimating the delay to completion of your works too early. Although it is not disastrous if you change your mind as events unfold, it certainly will not impress the contractor.

If you can include the particulars in your original notice so much the better (**document 7.03.4**). The particulars and estimate must be given separately for every item of delay which you think qualifies for extension of time. In other words, you must give the information for each just as though the other delay did not exist. This is not easy. Indeed, it is not easy to forecast the effect on the end date of a delay which occurs early in your programme of work. It is simplified if you have prepared a programme in the form of a network or precedence diagram at the beginning of the project (see section 5.01). There are now available several computer programs which assist this process and in seconds give the effect of delays either singly or cumulatively.

If you do find that passage of time or subsequent events cause you to revise your original estimates, you can inform the contractor accordingly (**document 7.03.5**). It is a requirement of all the sub-contracts that you do keep the contractor up to date.

Under the provisions of NSC/C, the architect has an important role. The contractor must pass on your original notice and any further particulars. The contractor may only give you an extension of time with the architect's written consent. The architect is only obliged to consider the matter on receipt of a request by the contractor and nominated sub-contractor under clause 35.14 of the main contract. In submitting particulars and estimate and any further notices, therefore, you must remember to request the contractor to join with you in requesting the architect's consent. If the contractor resists, it amounts to a breach of contract and you must tell him so (**document 7.03.6**).

7.03.1   Letter if delay occurs but no grounds for extension of time

To the Contractor

Dear Sirs,

[*Heading*]

The progress of the sub-contract works is being/is likely to be [*delete as appropriate*] delayed due to [*state reasons*].

We will continue to use our best endeavours to minimise the delay and its effects and we will inform you immediately the cause of the delay has ceased to operate.

This notice is issued in accordance with clause 11.2.1 [*substitute '12.2' when using NAM/SC or '2.2.1' when using NSC/C*].

Yours faithfully

7.03.2   Letter if delay gives grounds for extension of time

To the Contractor

Dear Sirs

[*Heading*]

It is apparent that progress/completion [*delete as appropriate*] of the sub-contract works is being/is likely to be [*delete as appropriate*] delayed due to [*state reasons*].   In our opinion, this is a matter which comes within clause 11.3.1 [*substitute '12.2.1 and/or 12.2.2' when using NAM/SC or '2.3.1' when using NSC/C*].

The delay began on [*insert date*].   When it is finished we will furnish you with our estimate of delay in the completion of the sub-contract works and further supporting particulars.

You may be assured that we are using our best endeavours to prevent delay in progress and completion of the sub-contract works.

This notice is issued in accordance with clause 11.2.1 [*substitute '12.2' when using NAM/SC or '2.2.1' when using NSC/C*].

Yours faithfully

7.03.3 Letter providing further particulars for extension of time

---

To the Contractor

Dear Sirs

[*Heading*]

We refer to our letter of the [*insert date*] in which we notified you
of delay in the progress of the sub-contract works likely to cause a
delay in the completion of the works. We note below particulars of
the expected effects and the estimated extent of delay in
completion of the sub-contract works in respect of each matter
specified in our notice:

[*List the matters separately, giving an assessment of the delay to
completion in each case, together with any supporting information*]

[*When using DOM/1, DOM/2 or NAM/SC add:*]

We believe that you now have sufficient information to enable you
to grant a fair and reasonable extension of time.

[*continued*]

7.03.3   *contd*

[*When using NSC/C add:*]

We believe that this is sufficient information to enable an extension of time to be granted and we request you under the provisions of clause 2.2.3 to submit the particulars and estimate to the architect and to join with us in requesting the consent of the architect under clause 35.14 of the main contract conditions.

Yours faithfully

7.03.4   Letter giving notice of delay and particulars in one notice

To the Contractor

Dear Sirs

[*Heading*]

It is apparent that progress/completion [*delete as appropriate*] of the sub-contract works is being/is likely to be [*delete as appropriate*] delayed due to [*state reasons*]. In our opinion this is a matter which comes within clause 11.3.1 [*substitute '12.2.2' when using NAM/SC or '2.3.1' when using NSC/C*].

We note below particulars of the expected effects and the estimated extent of delay in completion of the works in respect of each matter:
[*List each matter separately, giving an assessment of delay to completion in each case together with any other supporting information*]

[*When using DOM/1, DOM/2 or NAM/SC add:*]

We believe you now have sufficient information to enable you to grant a fair and reasonable extension of time.

[*continued*]

7.03.4   *contd*

[*When using NSC/C add:*]

We believe that this is sufficient information to enable an extension of time to be granted and we request you under the provisions of clause 2.2.3 to submit the particulars and estimate to the architect and to join with us in requesting the consent of the architect under clause 35.14 of the main contract conditions.

Yours faithfully

7.03.5 Letter updating the position regarding delays

To the Contractor

Dear Sirs

[*Heading*]

We refer to our letter of the [*insert date*] giving particulars and estimate of the effects of the delay referred to therein.

In accordance with the provisions of clause 11.2.2.3 [*omit the phrase completely when using NAM/SC or substitute '2.2.2.3' when using NSC/C*] we have to notify you that the situation has now developed as follows:

[*Explain precisely what changes have taken place, such as the continuance of a delay, a delay thought finished beginning again, the delay affecting different operations, etc.*]

[*When using DOM/1, DOM/2 or NAM/SC add:*]

Please take this information into consideration when giving an extension of time.

[*continued*]

7.03.5   *contd*

[*When using NSC/C add:*]

Please submit this information to the architect for his consideration in giving his consent under clause 35.14.

Yours faithfully

7.03.6 Letter if contractor will not join in requesting the architect's consent

This letter is only suitable for use with NSC/C
Registered post/recorded delivery

To the Contractor
(Copy to the Architect)

Dear Sirs

[*Heading*]

We refer to our letter of the [*insert date*] in which we gave notice of delay as required under the provisions of clause 2.2.1 and particulars and estimate as required under clause 2.2.2. We requested you to submit the particulars and estimate to the architect and to join with us in requesting the consent of the architect under clause 35.14 of the main contract. Our request was made as of right conferred on us by clause 2.2.3 of the sub-contract.

Your failure to comply with our request amounts to a breach of contract for which we shall be seeking appropriate damages unless you discontinue such breach within three days from the date of this letter and supply us with a copy of your letter to the architect.

Yours faithfully

7.04　Extension

The sub-contracts lay down differing periods within which the contractor must respond to your notices of delay. Under DOM/1 and DOM/2, clause 11.4, the contractor must respond within 16 weeks of receipt of the initial notice and reasonably sufficient particulars and estimates. NSC/C stipulates, in clause 2.3, that the response must be made within 12 weeks. In either case, if there is less than the stipulated time between receipt of sufficient particulars and the end of the sub-contract period, the contractor must respond before the end of the period.

NAM/SC, clause 12, does not set down any specific time period, merely requiring the contractor to act 'so soon as he is able to estimate' the delay. If the contractor does not act within the stipulated period, you should put his inaction on record (**document 7.04.1**). Sometimes, the contractor will attempt to delay the moment for his decision by requesting you to provide more and more information. When you have properly complied with the requirements of the contract, you can send him **document 7.04.2**. You must remember, however, that it is generally in your interests to obtain as great an extension of time as possible. Therefore, if you stick rigidly to your rights and supply only the minimum of information to the contractor, you will have little to complain about if he does not see your entitlement to extension of the sub-contract period in quite the same light as you do.

It is important to remember that the granting of an extension of time during the progress of the work is something of an interim affair. If, however, you are of the opinion that you have not received the extensions to which you are due, you should notify the contractor at once (**document 7.04.3**). In notifying you of the revised period for carrying out the sub-contract under DOM/1 and DOM/2, the contractor must state which of the relevant events he has taken into account and also the extent to which he has taken into account any omissions of work. NSC/C, of course, provides for the architect to agree these matters with the contractor before the contractor notifies you. NAM/SC makes no similar provision.

NSC/C provides, in clause 2.7, that if you feel aggrieved because the architect has not given his consent for an extension of time, not given it within the time limit or because of the content of the consent, you can require the contractor to let you use his name in arbitration proceedings (**document 7.04.4**).

Few contractors will observe the extension of time provisions to the letter. Usually, they will give you more information, indicating the thought process behind the extension. If you receive only the bald statement, you will have grave difficulty in raising a realistic objection, because in the case of the operation of numerous delaying factors, you may not know to which the contractor has apportioned which extensions. The best course of action, if the contractor will agree, is to meet him and get his explanation in detail. If he will not meet you, try **document 7.04.5**.

Each sub-contract provides for a review period after the date of practical completion of the sub-contract works. Under DOM/1 and DOM/2, clause 11.7, it must be carried out no later than 16 weeks after practical completion; under NSC/C, clause 2.5, the period is 12 weeks. NAM/SC makes similar provision in a very broadly drafted clause 12.4, by providing that the contractor may make an extension at any time.

DOM/1, DOM/2 and NSC/C provide for the contractor (with the architect's consent in the case of NSC/C) to make a decision to:

- confirm the current completion date; or
- fix a later date; or
- fix an earlier date if fair and reasonable having regard to any omissions since the last extension.

The review does not depend upon notice or on the existence of an extension of time for any particular relevant event. If the contractor fails to act, there will be no fixed contract period and the contractor's chances of recovering damages in respect of late completion of the sub-contract works will be correspondingly diminished. In the light of *Temloc Ltd* v. *Errill Properties Ltd* (1987) 39 BLR 30, the contractor may consider that he is not really bound to carry out the review within the time limits stated in the contract. The point is by no means settled as being of general application and you should waste no time in registering your contention that there is no fixed contract period (**document 7.04.6**). The contractor may not fix a date earlier than the completion date stated in the sub-contract documents. If he attempts to fix an earlier date, you must set the matter straight directly (**document 7.04.7**).

Each sub-contract has an important proviso that you must constantly use your best endeavours to prevent delay and do all reasonably required to the satisfaction of the contractor (and the architect in the case of NSC/C) to proceed with the works. This does not mean that you have to expend large sums of money, bring vastly increased resources onto the site nor comply with every direction without extra payment, although the contractor is unlikely to share that view.

Neither does it mean that the contractor has the power to direct you to accelerate the sub-contract works, with or without payment. If the contractor makes unreasonable demands, you should respond accordingly (**document 7.04.8**).

7.04.1 Letter if extension of time not granted within the time stipulated

To the Contractor
(Copy to Architect [when using NSC/C])

Dear Sirs

[Heading]

Notice of delay was sent to you on *[insert date]* in accordance with clause 11.2.1 *[substitute '12.2' when using NAM/SC or '2.2.1' when using NSC/C]* of the conditions of sub-contract. Full particulars including the effects of the delay and an estimate of the extent of delay in completion of the sub-contract works were sent to you on *[insert date]*. You made no request for further information.

[When using DOM/1, DOM/2 or NSC/C add:]

Clause 11.4.1 *[substitute '12.2' when using NAM/SC or '2.3' when using NSC/C]* requires you to respond within 16 *[substitute '12' when using NSC/C]* weeks of receipt of the notice, particulars and estimate. The period elapsed on *[insert date]* and you have not responded.

[continued]

[*When using NAM/SC add:*]

Clause 12.2 requires you to make a fair and reasonable extension
of the period for completion of the sub-contract works so soon as
you are able to estimate the length of delay beyond that period. It
is now at least three weeks since you could have carried out your
duty to so estimate and you have not made any extension of the
sub-contract period.

It is well established that failure to give an extension of time
properly due results in the sub-contract completion date becoming
inapplicable. In these circumstances our obligation is simply to
finish the sub-contract works within a reasonable time and the
provisions of clause 12 [*substitute '13' when using NAM/SC or
'clauses 2.8 and 2.9' when using NSC/C*] cannot be operated.

Yours faithfully

7.04.2   Letter if contractor unreasonably requests further information regarding delays

To the Contractor

Dear Sirs

[*Heading*]

Thank you for your letter of the [*insert date*] requesting further information before an extension of the sub-contract period can be given.

We submitted notice of delay, as required by the sub-contract, on [*insert date*]. We submitted full particulars including estimate of the effect of the delay on the completion date on [*insert date*]. We believe that you had all the information which could reasonably have been required by [*insert date*]. It is, of course, very much in our interests to supply you with full information as quickly as possible; this we have done.

It is our view that your latest request for the information is nothing but an attempt to postpone the giving of an extension of time. We, therefore, formally call upon you to carry out your duties under clause 11.3 [*substitute '12.2' when using NAM/SC or '2.3' when using NSC/C*] of the sub-contract conditions.

Yours faithfully

7.04.3   Letter if extension of time is insufficient

To the Contractor

Dear Sirs

[*Heading*]

We have received today your notification of an extension of time of [*insert period*] producing a new date for completion of [*insert date*].
We find your conclusions inexplicable in the light of the facts and the information we submitted in support of those facts.

Perhaps you would be good enough to reconsider your grant of extension of time or let us have an indication of your reasons for arriving at the time period you have granted. If you prefer, we would welcome a face to face discussion of the issues.

Yours faithfully

7.04.4 Letter if name borrowing arbitration required
This letter is only suitable for use with NSC/C
Registered post/recorded delivery

To the Contractor

Dear Sirs

[*Heading*]

We refer to our letters of the [*insert dates*] which was a notice under the terms of clause 2.2.1 and the particulars and estimate required by clause 2.2.2.

[*Add either:*]

The architect has failed to give the written consent referred to in clause 2.3

[*Or:*]

The architect has failed to give the written consent referred to in clause 2.3 within the period allowed in that clause

[*continued*]

7.04.4　*contd*

[*Or:*]

We note the terms of the architect's written consent dated　[*insert date*]
[*Then:*]

and we feel aggrieved thereby. We, therefore, formally call upon you to allow us to use your name and, as provided for under the terms of clause 2.7, join with us in arbitration proceedings to decide the matter as aforesaid. We should be pleased to receive your reply within seven days from the date of this letter.

Yours faithfully

7.04.5 Letter if contractor will not give detailed explanation of extension of time

To the Contractor

Dear Sirs

[*Heading*]

We refer to your notification of extension of time dated [*insert date*] and to our telephone conversation with Mr [*insert name*] of the [*insert date*].

We have made known our feelings that the length of the extension cannot possibly be justified by the facts. For your part, you continue to maintain that the extension you have given is fair and reasonable. You have refused to meet us to explain your reasoning and there appears to be a stalemate.

The obvious course of action is to refer the matter to arbitration. We are reluctant to do so because of the time and expense involved for all parties. It seems to us that, as reasonable people, we should be able to settle this matter without the need to resort to formal proceedings.

[*continued*]

7.04.5 *contd*

If you refuse to discuss the matter, we shall have no sensible alternative in view of what is at stake and if all else fails we will not shrink from that route. As an intermediate step, perhaps you will consider sending us a brief note of your reasons so that we can consider our position.

Yours faithfully

7.04.6   Letter if review of extensions not carried out
This letter is not suitable for use with NAM/SC

To the Contractor

Dear Sirs

[*Heading*]

Under the terms of the sub-contract clause 11.7 [*substitute '2.5' when using NSC/C*] you have a duty to carry out a review of the extensions of time and to either fix a new period for completion of the sub-contract works or to confirm the period previously fixed. The sub-contract stipulates that you must carry out this duty in writing no later than 16 [*substitute '12' when using NSC/C*] weeks from the date of practical completion of the sub-contract works.

You have failed to carry out your duty within the stipulated period. The result is that there is no fixed period for completion and our obligation is to finish within a reasonable time. This we have done.

Yours faithfully

7.04.7   Letter if attempt made to fix completion date earlier than date in the sub-contract

This letter is not suitable for use with NAM/SC

To the Contractor

Dear Sirs

[Heading]

We have received your letter of the *[insert date]* which purports to be given under the provisions of clause 11.7 *[substitute '2.5' when using NSC/C]*, fixing a revised date for completion of the sub-contract works of [insert date].

Under the provisions of the sub-contract Appendix, part 4 *[substitute 'NSC/T, Part 3, item 1' when using NSC/C]* the sub-contract completion date is *[insert date]*.

[When using DOM/1 or DOM/2 add:]

Clause 11.9 expressly provides that no decision under clauses 11.2 to 11.7 inclusive shall fix a period for completion of the sub-contract works shorter than the period stated in the Appendix, part 4.

[continued]

7.04.7 *contd*

[*When using NSC/C add:*]

It is a fundamental principle that you may not fix a period for completion of the sub-contract works shorter than the period in NSC/T, Part 3, item 1.

[*Then either:*]

There is no provision for you to carry out this duty more than once. Since your first exercise is clearly in breach of contract, no period has been fixed and our obligation is to complete within a reasonable time. This we have done.

[*Or:*]

We are now outside the period within which you are required to fix a new completion date. In any case you are not entitled to carry out this duty more than once. Since your first exercise was a clear breach of contract, no date has been fixed and our obligation is to complete within a reasonable time. This we have done.

Yours faithfully

7.04.8   Letter if the contractor alleges that best endeavours are not being employed

To the Contractor
(Copy to the Architect when using NSC/C)

Dear Sirs

[Heading]

Clause 11.8 *[substitute '12.5' when using NSM/SC or '2.5' when using NSC/C]* of the conditions of sub-contract requires us to use constantly our best endeavours to prevent delay. Your allegation in *[state where and date, e.g. minute no. 5.3, Progress Meeting of the 18 July 1994]* that we are failing to carry out our duties in this respect is totally without foundation.

Our obligation to use best endeavours is simply an obligation to continue to work regularly and diligently, re-arranging our labour force as best we can. This we have done and we are continuing so to do, responding to your reasonable requests in the spirit of the sub-contract. There is no obligation upon us to spend additional sums of money to make up lost time.

If you purport to take any action or fail to take some action on the grounds of our alleged failure to use best endeavours, we will take immediate advice on the remedies available to us.

Yours faithfully

7.05 Failure to complete on time

Clause 12 of DOM/1 and DOM/2, clause 13 of NAM/SC and clause 2.8 of NSC/C deal with the situation if you fail to complete the sub-contract works within the period allowed. Under DOM/1 and DOM/2, the contractor must give you written notice, within a reasonable time of the expiry of the period, stating that you have failed to complete. This notice is important because it triggers the next stage in the clause. When you receive the notice, you must pay or allow to the contractor a sum representing whatever loss or damage he has suffered as a result of your failure to complete. In practice, the contractor will invariably set off the amount in question from your next payment. But without such a notice, he would be unable to set off lawfully on account of your late completion.

On receipt of such a notice, therefore, you must waste no time in challenging it if you think it is wrong (**document 7.05.1**). Grounds for challenge might well be that you actually finished at the right time, or that there is no longer a fixed period for completion, or that you have not been given all the extensions of time to which you are entitled, or that the contractor did not give the notice within a reasonable time of the expiry of the period and so on. What is a reasonable time will vary depending upon circumstances. In practice, it will probably be interpreted liberally, particularly if it can be said that you are well aware of the situation and of the fact that the contractor is suffering loss or damage as a result of your shortcomings.

The most common form of loss or damage suffered by the contractor will be the deduction of liquidated damages under the main contract. In order to be able to deduct them from you he must show that, for example, the fact that you are two weeks late has delayed the main contract by two weeks. The reality is that he will usually deduct and leave you to show reasons why he should not do so (see section 6.05).

Under NSC/C, clauses 2.8 and 2.9, the position is very similar, but there are some important differences. If you fail to complete the sub-contract works within the stipulated period, the contractor must notify the architect, giving you a copy of the notification. Under clause 35.15 of the main contract (JCT 80), it is for the architect to certify in writing to the contractor, with a copy to you, that you have failed to complete. Before he does so, however, the following criteria must be satisfied:

- The contractor must have notified the architect.
- The contractor must have sent you a copy of the notice.

- The architect must be satisfied that the extension of time clause has been properly operated.

The architect must give any certificate under clause 35.15 within two months of the date of the contractor's notice that you have failed to complete the sub-contract works. You have scope for challenging the certificate on the grounds that you have not received the extensions of time to which you are due, or that you did not receive a copy of the contractor's notice, or that the architect's certificate was not issued within the two month period (**document 7.05.2**). It is worth making your objection known immediately the contractor sends his initial notice to the architect. The architect will naturally be anxious that in giving a certificate under clause 35.15 of the main contract he is being scrupulously fair. An appropriate letter at this point, with a copy to the architect, may persuade him that he should not issue a certificate at all (**document 7.05.3**).

The architect's certificate is a precondition for the contractor to claim against you on the grounds of late completion. If he attempts to claim from you without such a certificate, you must respond accordingly (**document 7.05.4**).

NAM/SC, clause 13, follows a similar pattern to DOM/1 in that the contractor must notify you in writing within a reasonable time if you fail to complete. If he gives you a further extension of time after the issue of the notice, the notice is deemed cancelled and, presumably, he must give you another. Clause 13, however, gives no rights to the contractor to claim from you in respect of late completion. He has, of course, his rights under clause 14.3 to claim for disturbance of regular progress and his rights of set-off under clause 21.

7.05.1 Letter if contractor gives notice of non-completion
This letter is not suitable for use with NSC/C

To the Contractor

Dear Sirs

[*Heading*]

We are in receipt of your letter of the [*insert date*] which purports to be a notice under the provisions of clause 12.1 [*substitute '13' when using NAM/SC*]. In our view the notice is invalid because

[*Either:*]

we completed the sub-contract works within the period for completion
[*Or:*]

for the reasons already stated in our letter of the [*insert date*], our obligation is to complete within a reasonable time and we have discharged that obligation. If you contend to the contrary, we will put you to proof, but in any event clause 12 [*substitute '13' when using NAM/SC*] does not provide for such a situation.

[*continued*]

7.05.1   *contd*

[*Or:*]

we have not been given the extensions of time to which we are
entitled.
[*Or:*]

you have failed to comply with the terms of clause 12.1 [*substitute
'13' when using NAM/SC*].

[*Then:*]

We, therefore, formally request you to withdraw your purported
notice. Any attempt to deduct money from payments due to us will
be met with legal action.

Yours faithfully

7.05.2 Letter if architect gives certificate of non-completion
This letter is only suitable for use with NSC/C
Registered post/recorded delivery

To the Contractor
(Copy to Architect)

Dear Sirs

[*Heading*]

We are in receipt of a copy of the architect's certificate dated
[*insert date*] certifying that we have failed to complete the
sub-contract works within the period provided. The certificate is
not validly given because
[*Either:*]

you have failed to comply with our request under clause 2.2.3.

[*Or:*]

we have not been given the extensions of time to which we are
due.

[*continued*]

7.05.2 *contd*

[*Or:*]

we have not received a copy of your notice to the architect as provided under clause 2.8. The giving of a copy to us is clearly a condition precedent to the architect issuing his certificate under clause 35.15.1 of the main contract.

[*Or:*]

your notice under clause 2.8 was given on [*insert date*]. The architect's certificate is dated [*insert date*]. The difference is more than the two months allowed under clause 35.15.2 of the main contract.

[*Then:*]

We hereby give formal notice that the certificate should be withdrawn. Any action taken on the basis of the architect's certificate will be firmly resisted.

Yours faithfully

7.05.3   Letter if the contractor wrongly sends a clause 2.8 notice to the architect
This letter is only suitable for use with NSC/C

To the Contractor
(Copy to Architect)

Dear Sirs

[*Heading*]

We have received a copy of your letter to the architect dated
[*insert date*] which purports to be a notice under the provisions of
clause 2.8 of the sub-contract. Your notice is not validly given
under clause 2.8 and, therefore, the architect cannot issue a
certificate under clause 35.15.1 of the main contract because

[*Either:*]

you have failed to comply with our request under clause 2.2.3.

[*Or:*]

we have not been given the extensions of time to which we are
due.

[*continued*]

7.05.3   *contd*

[*Or:*]

for the reasons already given in our letter of the [*insert date*] our obligation is to complete the sub-contract works within a reasonable time. We have properly discharged that obligation.

[*Then:*]

Your notice, therefore, is in breach of contract and it should be withdrawn forthwith.

Yours faithfully

7.05.4 Letter if the contractor attempts to claim without an architect's certificate
This letter is only suitable for use with NSC/C

To the Contractor

Dear Sirs

[*Heading*]

We have received your letter of the [*insert date*] in which you state that you intend to claim a sum equivalent to the loss or damage you have suffered or incurred as a result of our alleged failure to complete within the period for completion of the sub-contract works.

Without prejudice to our right to challenge your assertion that we have so failed, we draw your attention to clause 2.9 of the sub-contract which clearly stipulates that you shall not be entitled so to claim unless the architect in accordance with clause 35.15 of the main contract shall have issued to you a certificate in writing with a copy to us certifying any failure notified under clause 2.8 of the sub-contract.

The architect has not issued a certificate under clause 35.15 of the main contract or if he has, he has not, as he is contractually bound to do, issued a copy to us. Neither do we believe that the architect has proper grounds for issuing such a certificate. Therefore, should you put your claim into effect, we will take immediate legal action to recover such sums and damages for breach of contract.

Yours faithfully

Chapter 8

Claims and Counterclaims

8.01 Direct loss and/or expense

The contractual machinery for reimbursement of direct loss and/or expense is contained in clause 13 in DOM/1 and DOM/2, clause 14 in NAM/SC and clauses 4.38 to 4.41 in NSC/C. In each case the clause deals, in separate sections, with your monetary claims against the contractor and his claims against you.

The provisions in each contract are different, but NSC/C demonstrates the most significant differences which are a consequence of the architect's involvement in the ascertainment process. In this form, there are in fact three sections: your claims against the contractor for matters which are the responsibility of the employer, your claims for matters which are the contractor's own responsibility, and his claims against you.

It is important to remember that the contract sets out a procedure by which loss and/or expense can be recovered in respect of certain occurrences provided you or the contractor follows the rules laid down. The contract does not pretend to cover exhaustively all the occurrences which either you or the contractor may decide entitle you to recover loss. If you do not fulfil the conditions in the claims clause, you can still pursue a claim under the general law in arbitration or litigation, provided of course that the event giving rise to the claim amounts to a breach of contract.

Just as not all possible matters are catered for in the claims clause, so you will see that many of the matters for which you can obtain reimbursement through the contractual machinery are not matters which would entitle you to damages at common law. Take opening up and testing for example. Each contract provides that if you are instructed to open up part of the sub-contract works for inspection and if the work is found to have been carried out in accordance with the contract, you are entitled to make an application for any loss and/or expense you have suffered. Clearly an instruction of that kind is empowered under the terms of the sub-contract and, therefore, it cannot be a breach of that contract. In those circumstances, you would have no claim against the contractor through the ordinary legal process although the contract does provide you with the opportunity to recover any direct loss and/or expense which you have incurred.

'Direct loss and/or expense' means that what you can recover are those heads of claim which would be recoverable in an action for damages for breach of contract at common law. The word 'direct' puts a limit on the damage which is recoverable. The loss must be a direct result of the matter referred to and cannot be loss which you might say would not have been incurred 'but for' the matter.

For example, if you are instructed to carry out extra works, you would be entitled to claim for the cost of the disruption or prolongation of the sub-contract works as a result of the extra work. This might entail claiming, among other things, for the additional cost of supervision for an extended period or difficult working arrangements which are a clear result of the instructions issued at that time. All that would be 'direct' loss. If, however, a piece of equipment which you had to bring onto site specifically to carry out the extra work broke down, the costs involved during the period it was out of commission would not be recoverable because although it might be possible to say that they would not have been incurred 'but for' the instruction, it is clear that the cause of the loss was insufficient maintenance, not the instruction.

8.02 Procedure: sub-contractor's claims against the contractor
In each case, the procedure is clearly laid down, but the provision repays careful reading. Many of the conditions have a common basis although they may be expressed differently. NAM/SC and NSC/C make clear that the loss and/or expense must not be capable of reimbursement under any other provision of the contract. Although not expressly stated in DOM/1 or DOM/2, it would be implied as a matter of both common sense and common law.

The word 'claim' is not mentioned; instead, you must make application to the contractor in writing (**document 8.02.1**) as soon as it has become or should reasonably have become apparent that regular progress has been or is likely to be affected. Within five lines, DOM/1 and DOM/2 stipulate that the application must be made 'as soon as' and 'within a reasonable time of'. NAM/SC has a variation in that the application should be made within a reasonable time of it becoming apparent that you have incurred or are likely to incur loss and/or expense. The regular progress of the sub-contract works must have been materially affected. That is to say that the progress must have been affected substantially. The clause is not intended to cater for trivial disruptions.

It is safest to make the application as soon as you can, even if you may not be fully aware of the total consequences of an occurrence. Under DOM/1, DOM/2 and NAM/SC, the amount of loss and or expense is to be agreed between you. You are obliged to supply the contractor with whatever further information the contractor, or in the case of NSC/C the architect, may require in support of your application, as is reasonably necessary to show that regular progress has been materially affected, or reasonably to enable ascertainment to take place (**document 8.02.2**). It is usually wise to submit a very simple application with basic supporting evidence. It should be unnecessary to prepare or to have someone else prepare for you a very elaborate set of claim documents. Neither the contractor nor the architect are strangers to the site. If you consider that their demands are unreasonable you should put it in writing, if only for record purposes (**document 8.02.3**). Bear in mind, however, that if you refuse to provide what the contractor maintains is reasonable information, he may make that an excuse for delaying the agreement of your entitlement or for ascertaining far less than is your due. In some instances nothing less than a fully worked out claim document will suffice to demonstrate your entitlement.

DOM/1, DOM/2 and NAM/SC refer to the agreed amount of loss and/or expense. In practice, of course, the contractor is not going to be happy to agree to more than he ascertains is due. Under the provisions of NSC/C, the contractor must require the architect to operate the main contract clause 26.4 so that the amount of loss and/or expense may be ascertained. If you discover that he has not done so you must notify him immediately with a copy to the architect (**document 8.02.4**).

NSC/C has a separate section, clause 4.39, which deals with your claims against the contractor in regard to matters which are his own responsibility – any act, omission or default. Each contract has that provision, but only NSC/C has to separate it because it is not something with which the architect will be concerned. In other respects, the clause is in familiar terms. It hardly needs saying that a contractor will be less inclined to look favourably on this type of claim because he has little or no chance of passing it up to the employer.

Many contractors are not prepared to look at a financial claim unless an extension of time has been given for the same reason. Likewise, many sub-contractors think that they are not entitled to reimbursement unless an extension of time has been given. There is no foundation in the contract or elsewhere for this view. Indeed, case law confirms that reimbursement of loss and/or expense does not depend on extension of the sub-contract period. You may have a perfectly valid entitlement to loss and/or expense whether you finish on time, earlier or later than the sub-contract period. You should strongly resist any attempt on the part of the contractor to limit your entitlement (**documents 8.02.5 and 8.02.6**).

NSC/C clearly refers in clause 4.38.3 to 'any amount from time to time ascertained'. The philosophy behind the clause is that ascertainments will be made as applications are made, not left until the end of the sub-contract or even the main contract period. The other forms do not make express reference to ascertainment from time to time, but it would probably be implied. If the contractor seeks to postpone consideration of your application, you should object (**document 8.02.7**). **Document 8.02.8** is a similar letter if the problem is that the architect will not ascertain.

If the problem is that the contractor will not agree the amount you think is your due, you should register your objection very strongly (**document 8.02.9**). Under NSC/C, if you are dissatisfied with the architect's ascertainment, you should write to the contractor with a copy to the architect (**document 8.02.10**).

Many sub-contractors are doubtful about writing to the architect or copying letters to him. Architects often take the line that they are not concerned with the sub-contract. Both views are understandable up to a point, but where the architect is specifically given a role in relation to the sub-contract, or where some benefit to the sub-contractor depends directly on the architect's action, there is every reason for the sub-contractor, if necessary, to acquaint the architect directly about certain matters. Sadly, the architect may sometimes respond rather brusquely. There is no excuse for this kind of behaviour and although you will be mindful of the need to avoid antagonising the architect, it should not prevent you from asserting your rights (**document 8.02.11**).

If all else fails, you may be obliged to pursue your claim in arbitration. Whether you do this will inevitably be a commercial decision after taking every consideration into account.

8.02.1   Letter making application for loss and/or expense

To the Contractor

Dear Sirs

[*Heading*]

In accordance with clause 13.1 [*substitute '14.1' when using NAM/SC or '4.38.1' when using NSC/C*] we hereby make application that we are likely to incur direct loss and/or expense and financing charges in the execution of this sub-contract, for which we will not be reimbursed by a payment under any other provision, because the regular progress of the sub-contract works is likely to be materially affected by [*insert description*] being a matter under clause [*note one of clauses 13.3.1-7 when using DOM/1 or DOM/2, or 14.2.1-9 when using NAM/SC, or 4.38.2.1-8 when using NSC/C*].

[*When using NSC/C add:*]

Please require the architect to operate clause 26.4 of the main contract conditions so that the amount of that direct loss and/or expense may be ascertained.

Yours faithfully

8.02.2 Letter submitting further information

To the Contractor

Dear Sirs

[*Heading*]

Thank you for your letter of the [*insert date*] requesting further information in support of our application for direct loss and/or expense sent to you on the [*insert date*].

[*Give a detailed description of the circumstances in a factual way and include all the evidence you have in support. This may be such things as progress reports, marked up programmes, particularly if they are of the network variety, correspondence, invoices, etc.*]

We consider that this is sufficient information to enable you to agree [*substitute 'the architect to ascertain' when using NSC/C*] the amount to which we are due.

Yours faithfully

8.02.3   Letter if unreasonable demands for further information are made

To the Contractor

Dear Sirs

[*Heading*]

Thank you for your letter of the [*insert date*] in which you request us to provide more information in respect of our application for direct loss and/or expense first made on [*insert date*]. We have already responded to requests for further information on [*insert dates*] and we consider that your latest demands are unreasonable [*add if appropriate: 'and almost impossible to fulfil'*]. We enclose, however, such of the information requested as we can reasonably put together and we now formally request you to agree with us the amount of direct loss and/or expense to which we are entitled.

Yours faithfully

8.02.4   Letter if contractor fails to require the architect to operate clause 26.4 of the main contract

This letter is only suitable for use with NSC/C
Registered post/recorded delivery

To the Contractor
(Copy to Architect)

Dear Sirs

[*Heading*]

We sent you our application under the provisions of clause 4.38.1 on the [*insert date*]. You have a duty under the clause to require the architect to operate clause 26.4 of the main contract conditions so that the amount of the direct loss and/or expense may be ascertained. We understand that you have not yet carried out your obligations in this respect.

Financing charges are being incurred which we will seek to recover either as part of the direct loss and/or expense or part of damages for your breach of the sub contract.

We now formally require you to carry out your duty under clause 4.38.1 of the sub-contract within seven days of the date of this letter.

Yours faithfully

8.02.5 Letter if contractor seeks to limit loss and expense to extensions of time given

This letter is not suitable for use with NSC/C

To the Contractor

Dear Sirs

[*Heading*]

We are in receipt of your letter of the [*insert date*] agreeing to an amount of direct loss and expense which you seek to confine to amounts in respect of the extended sub-contract period for which we have received an extension of time.

There is nothing in clause 13 [substitute '14' when using NAM/SC] which limits our entitlement in this way. The clause quite clearly entitles us to all loss and/or expense directly incurred due to the material effect on regular progress of the sub-contract works arising from [*state the matters*].

We request you to reconsider your letter in the light of our comments and the sub-contract terms and notify us of the amount you are prepared to agree which realistically reflects the true situation.

Yours faithfully

8.02.6 Letter if architect seeks to limit loss and expense to extensions of time given

This letter is only suitable for use with NSC/C

To the Contractor
(Copy to Architect)

Dear Sirs

[*Heading*]

We are in receipt of your letter of the [*insert date*] from which we understand that the architect is seeking to confine the amounts of direct loss and expense to amounts in respect of the extended sub-contract period for which we have received an extension of time.

There is nothing in clause 4.38 which limits our entitlement in this way. The clause quite clearly entitles us to all loss and/or expense directly incurred due to the material effect on regular progress of the sub-contract works arising from [*state the matters*].

We, therefore, request you to refer this matter back to the architect and require him to correctly operate the provisions of clause 26.4 of the main contract conditions. Should he fail to do so we shall seek immediate advice with a view to operating appropriate legal and/or contractual remedies.

Yours faithfully

8.02.7   Letter if agreement is delayed
This letter is not suitable for use with NSC/C

To the Contractor

Dear Sirs

[*Heading*]

We refer to our application of the [*insert date*] submitted in
accordance with the provisions of clause 13.1 [*substitute '14.2.1'
when using NAM/SC*] and the supporting information submitted
on [*insert dates*].

[*Insert number*] weeks have elapsed since we last submitted such
information to you and no further details have been requested.
Indeed, we believe that you have been given everything required to
enable agreement to be reached in respect of the loss and expense
we have incurred. The sub-contract does not envisage that
agreement will be delayed until the end of the sub-contract period.
On the contrary, the clear scheme is that amounts due under clause
13 [*substitute '14' when using NAM/SC*] will be agreed and paid
as the sub-contract works proceed.

We look forward to receiving your positive response with seven
days from the date of this letter, failing which we will take
appropriate action.

Yours faithfully

8.02.8   Letter if ascertainment is delayed
This letter is only suitable for use with NSC/C

To the Contractor
(Copy to Architect)

Dear Sirs

[*Heading*]

We refer to our application of the [*insert date*] submitted in accordance with the provisions of clause 4.38.1 and the supporting information submitted on [*insert dates*] as requested by the architect.

[*Insert number*] weeks have elapsed since we last submitted such information to you and we must assume that you passed the information to the architect. The architect has not, to our knowledge, requested you to obtain further information from us. Indeed, we believe that we have given you everything required to enable ascertainment to be carried out. Clause 4.38.3 of the sub-contract clearly states that 'any amount from time to time ascertained' must be added to the sub-contract sum. The scheme is that amounts of loss and expense will be ascertained and paid as the sub-contract works proceed.

It is now reasonable for us to request you to require the architect to operate the provisions of main contract clause 26.4 forthwith, failing which we will pursue our contractual and/or legal remedies.

Yours faithfully

8.02.9   Letter if the contractor will not agree the amount due
This letter is not suitable for use with NSC/C

To the Contractor

Dear Sirs

[*Heading*]

We refer to our application dated [*insert date*] in respect of loss
and expense. Your letter of the [*insert date*] notifying us of the
amount you are prepared to agree appears to take little account of
the very full supporting information submitted by us on [*insert
dates*].

Unless we hear from you by [*insert date*] that you will amend the
amount you are prepared to agree as the amount of our loss and
expense to take account of the information we have supplied, we
will take appropriate steps to refer this dispute to arbitration in due
course.

Yours faithfully

8.02.10   Letter if dissatisfied with the architect's ascertainment
This letter is only suitable for use with NSC/C

To the Contractor
(Copy to Architect)

Dear Sirs

[*Heading*]

We refer to our application dated [*insert date*] in respect of loss and expense. The amount ascertained by the architect, as notified to us by your letter of the [*insert date*] appears to take little account of the very full supporting information submitted by us on [*insert dates*].

Unless we hear from you by [*insert date*] that the architect will amend his ascertainment to take account of the information we have supplied, we will take the appropriate steps to refer this dispute to arbitration in due course.

Yours faithfully

8.02.11   Letter if the architect responds brusquely to sub-contractor's approach

---

To the Architect

Dear Sir

[*Heading*]

Thank you for your letter of the [*insert date*]. We regret that you have replied in those terms. There is absolutely no reason why we should not write to you directly about sub-contract matters which depend on your actions. It is customary and indeed sensible that we should normally communicate with the main contractor who is the other party to our sub-contract, and the sub-contract terms clearly envisage that this will be the normal channel for correspondence, notices, etc.   There will be occasions however, and this we believe is such an occasion, when it is most appropriate for us to write directly to you.

We hope that you will now properly respond to our earlier letter. If you do not, we are left with no real alternative but to take legal proceedings to secure our rights.

Yours faithfully

8.03 Contractors' claims against the sub-contractor

Contractors' claims under DOM/1, DOM/2 and NAM/SC are similar in effect although in somewhat differing formats in clauses 13.4 and 14.3 respectively. If regular progress of the works (not simply the sub-contract works) is substantially affected by any of your acts, omissions or defaults or those of any of your servants or agents, and if the contractor wishes to recover any direct loss and/or expense he has suffered, he is to make written application within a reasonable time of the effect becoming apparent. The agreed amount of loss and/or expense may be deducted from any money due to you or it may be recovered from you as a debt. In other words, the contractor may set off the amount or, if there is not sufficient money still to be paid, he may sue for it through the courts.

If, of course, you do not agree, he cannot act under this clause although he may still take action under the set-off provisions if he can bring your act, omission or default under those provisions as constituting a breach of the sub-contract or a failure to observe its terms. You must respond to the contractor's application and if you do not agree, make the point abundantly clear (**document 8.03.1**). There is an important proviso that the contractor must provide, on your request, such information as is reasonably necessary to enable agreement to take place and, indeed, to establish in the first place that regular progress was indeed disrupted.

If, as is often the case, the contractor simply sends a bald application for the amount he considers is due, you must immediately press him for further information (**document 8.03.2**). If the further information is insufficient, you must press him again (**document 8.03.3**). In order to show that regular progress has been disrupted, it is not enough simply for the contractor to show that he has been prevented from working to his programme. The programme is purely an indication of his intentions unless, as noted earlier, it is a contract document, which is rare. He must show that:

- Actually, not just in theory, his regular progress has been disrupted.
- The disruption is directly resulting from your act, omission or default.
- The loss and/or expense is directly caused by the disruption.

Needless to say, in practice it is not easy to make these important connections even though the standard of proof required is the balance of probabilities. If the contractor fails to produce evidence which makes his case, you must tell him (**document 8.03.4**).

The position under NSC/C clause 4.40 is broadly similar, but there are some significant differences. It is expressly stated that the 'Works' includes any part which is sub-sub-contracted. Later, the clause refers to loss and/or expense caused to the contractor and makes clear that it includes loss and/or expense suffered by other sub-contractors. There is an all important proviso that the claims from other sub-contractors must have been agreed by the contractor, the other sub-contractors and you. Moreover, the claims of other sub-contractors must have been made under 'similar provisions in the relevant sub-contracts'. It is nowhere stipulated to what the provisions must be similar. Presumably, it refers to the clause in which the phrase is found. There will, of course, be similar provisions in other sub-contracts, but they will not enable another sub-contractor to claim because the clause is expressly for the benefit of the contractor and his claims.

This seems to be a major stumbling block to recovery of claims by the contractor of other sub-contractors' claims against him in respect of disruption of regular progress as a result of your act, omission or default. We venture to suggest that reference to other sub-contractors might have been better omitted because recovery of such losses to the contractor would in any case be implied. The insertion of the reference in this instance appears to limit rather than extend the scope of this clause.

8.03.1   Letter if contractor's claim not valid

To the Contractor

Dear Sirs

[Heading]

We refer to your letter of the *[insert date]* which purports to be an application under clause 13.4 *[substitute '14.3' when using NAM/SC or '4.40' when using NSC/C]*.

We do not consider that the occurrence to which you refer is an act, omission or default attributable to us or to anyone for whom we are responsible, neither did the alleged occurrence materially affect the regular progress of the works. It, therefore, follows that we cannot agree the sum you propose as your loss and expense which would be excessive even if your claim was well founded.

Yours faithfully

8.03.2   Letter requesting further information

To the Contractor

Dear Sirs

[*Heading*]

We refer to your application for loss and/or expense dated [*insert date*].

In order reasonably to enable us to operate the provisions of clause 13.4 [*substitute '14.3' when using NAM/SC or '4.40' when using NSC/C*] of the sub-contract, we should be pleased to receive the following information:

[*List the information required*]

Yours faithfully

8.03.3   Letter if further information is not sufficient

To the Contractor

Dear Sirs

[*Heading*]

Thank you for your letter dated [*insert date*] referring to your
claim for loss and/or expense in response to our request for further
information reasonably necessary for the purposes of clause 13.4
[*substitute '14.3' when using NAM/SC or '4.40' when using NSC/C*].

You have not produced the information required and until you do,
we can make no further progress on this matter.

Yours faithfully

8.03.4   Letter if contractor does not establish his entitlement to loss and/or expense

To the Contractor

Dear Sirs

[*Heading*]

Thank you for your letter of the [*insert date*] in connection with your application for loss and/or expense dated [*insert date*].

We have carefully considered all the information you have submitted, together with what we know of the works. We cannot agree any amount of loss and/or expense due to you because you have failed to produce evidence to support your contentions in regard to the act, omission or default you allege, you have failed to show that any such alleged acts, etc. have materially affected the works and you have failed to show that the amount you specify was a direct result of the foregoing.

Yours faithfully

Chapter 9

Completion and Defects Liability

9.01 Practical completion

Practical completion of the sub-contract works is referred to in each of the sub-contracts under consideration. DOM/1, DOM/2 and NAM/SC actually also refer to the sub-contract works being 'practically completed', but it is not thought that the distinction is of any significance. 'Practical completion' is not defined anywhere in the documents and there is a difference of view regarding when practical completion can be said to have occurred in any particular instance. There have been judicial pronouncements which, although not entirely consistent, seem to suggest that practical completion may be said to have occurred when the work is complete except for very minor items still to be completed and provided that there are no obvious (patent) defects.

The date of practical completion is important because, among other things, it marks the end of the sub-contractor's liability for damages as a result of failure to complete within the sub-contract period and it signals the release of half the retention money. You will be anxious that practical completion occurs as early as possible.

The procedure is dealt with in DOM/1 and DOM/2 clause 14, NAM/SC clause 15 and NSC/C clauses 2.10 to 2.14. NSC/C is the only form which provides for practical completion to be certified by the architect. It is more convenient to consider those terms separately.

In DOM/1, DOM/2 and NAM/SC, you are to notify the contractor in writing when, in your opinion, the sub-contract works are practically completed (**document 9.01.1**). This is in marked contrast to the main contract forms which leave it to the architect to initiate the process. The contractor has fourteen days in which to dissent from your notice in writing. If he does not so dissent, practical completion is deemed to have taken place for all the purposes of the sub-contract (for example, release of half the retention) on the date you have notified.

A very important proviso states that the contractor must state his reasons for dissenting. He cannot, therefore, merely state that he dissents. Such a dissent would not be valid and it is likely that your notice of practical completion would prevail (**document 9.01.2**). The contractor's reasons for dissent are not stipulated to be reasonable, but in practice unreasonable dissent would defeat the whole object of the clause, which is to fix the date on which practical completion actually occurred. If the contractor does so dissent with reasons, no doubt you will immediately take issue with him (**document 9.01.3**).

Unless the contractor withdraws his dissent after your protest, practical completion will then be deemed to have taken place on such date as may be agreed. It is easy to visualise many instances when agreement cannot be reached. In such circumstances, there is a nasty sting in the tail of this clause which, in the event of failure to agree, deems practical completion taking place on the date of practical completion of the works as certified by the architect under the main contract. If you are a sub-contractor whose work is normally complete long before practical completion of the main contract, this could be disastrous in terms of retention and damages for delay. In practice, damages for delay under the sub-contracts must be proved, unlike the main contract provisions for liquidated and ascertained damages, and this requirement should ease the situation considerably. In the case of, for example, a piling sub-contractor, the provision is clearly ludicrous, because his work must be complete before other work can proceed. If, therefore, the contractor fails to agree and maintains that practical completion is not deemed to have taken place until practical completion of the main contract works, you must register the facts strongly (**document 9.01.4**). Alternatively, of course, you can seek arbitration, but it is probably best to try sweet reason first.

The position under NSC/C differs to take account of the fact that practical completion is to be certified by the architect. Once again, the process is triggered by your notice to the contractor (see **document 9.01.1**). The contractor must pass your notice to the architect immediately, together with any observation which the contractor may wish to make. The contractor must send a copy of those observations to you.

Three situations can arise. First, the contractor may not pass your notice to the architect at all. You will not be aware of this until it becomes clear that the architect has not issued his certificate. He must do this forthwith and send you a copy if he is of the opinion that practical completion has taken place.

Second, the contractor may not send you a copy of his observations. As far as you are concerned, this situation will be rather like the first in that you will have served notice on the contractor and you will have heard nothing further. A single letter, copied to the architect, will serve both purposes (**document 9.01.5**). You will note that it is couched in the form of an enquiry about the safe receipt of your notice and commenting that you assume the contractor has no adverse observations. On receipt of your copy, the architect will be put on notice that he should be operating clause 35.16 of the main contract and certifying practical completion of the sub-contract works.

The third situation is that the contractor may send you a copy of his observations which may recommend to the architect that practical completion has not taken place. Clearly, you will disagree with these sentiments and you must notify the contractor without delay, once again sending a copy to the architect (**document 9.01.6**).

If the architect issues his certificate under clause 35.16 of the main contract and you disagree, because, for example, it specifies a date two weeks later than the date you consider to be correct, you can take the matter to immediate arbitration. Whether or not you do so will largely depend on how much is at stake financially. There is no point in seeking arbitration purely on principle. Before you seek arbitration, you might try a letter to the contractor with a copy to the architect, setting out the true position as you see it (**document 9.01.7**).

9.01.1   Letter giving notice of practical completion
Registered post/recorded delivery

To the Contractor

Dear Sirs

[Heading]

This is the notice under clause 14 *[substitute '15' when using NAM/SC or '2.10' when using NSC/C]* of the sub-contract conditions that in our opinion the sub-contract works are practically complete *[substitute 'will have reached practical completion' when using NSC/C]* on *[insert date]*.

[When using NSC/C add:]

We should be pleased if you would now operate the provisions of clause 2.10 and immediately pass this notification to the architect so that he can issue a certificate of practical completion of the sub-contract works under clause 35.16 of the main contract.

Yours faithfully

9.01.2   Letter if the contractor merely dissents
This letter is not suitable for use with NSC/C

To the Contractor

Dear Sirs

[*Heading*]

We are in receipt of your letter of the [*insert date*] which purports
to be a notice of dissent to our notice that the sub-contract works
are practically completed which we sent to you on [*insert date*].

Clause 14 [*substitute '15' when using NAM/SC*] expressly states
that your written notice of dissent must set out your reasons for
such dissent. You have clearly failed to set out any reasons at all.
Your letter, therefore, is not a valid notice of dissent. It is merely
vexatious. Another very important provision of clause 14
[*substitute '15' when using NAM/SC*] states that you must dissent
within 14 days from receipt of our notice. At the date of this letter
it is [*insert number*] days since you received our notice. You have
not served a valid notice of dissent on us and, therefore, the date
of practical completion is [*insert date*] for all the purposes of the
sub-contract.

Yours faithfully

9.01.3   Letter if you disagree with contractor's reasons for dissent
This letter is not suitable for use with NSC/C

To the Contractor

Dear Sirs

[*Heading*]

We are in receipt of your letter of the [*insert date*] in which you state that you do not agree with the date of practical completion which we have correctly stated as [*insert date*].

We note the reasons which you give for dissent and we must state quite plainly that we consider your reasons wrong/misguided/ unfounded/vexatious [*delete as appropriate*], because [*insert your reasons for disagreeing*].

We, therefore, request you to withdraw your dissent. If you fail to so withdraw, we will take whatever legal steps are necessary to protect our interests.

Yours faithfully

9.01.4 Letter if contractor will not agree the date of practical completion

This letter is not suitable for use with NSC/C

To the Contractor

Dear Sirs

[*Heading*]

Thank you for your letter of the [*insert date*] from which we note that you are not prepared to agree the date of practical completion of the sub-contract works. You now contend that, because you have prevented agreement, practical completion of the sub-contract works must be deemed to have taken place on the date of practical completion of the works as certified by the architect under clause 17.1 [*substitute '2.9' when using NAM/SC*].

Your stance is demonstrably obstructive and your contention cannot be sustained in the light of the facts. If, therefore, you do not withdraw your contention and give clear indications of willingness to agree a realistic date, we shall have no alternative but to seek immediate arbitration.

Yours faithfully

9.01.5   Letter if there is no response from architect or contractor
This letter is only suitable for use with NSC/C

To the Contractor
(Copy to Architect)

Dear Sirs

[*Heading*]

We refer to our letter of the [*insert date*] which was a notice
under clause 2.10 of the sub-contract conditions that in our opinion
the sub-contract works would have reached practical completion on
[*insert date*].

The notice was sent by registered post/recorded delivery [*delete as
appropriate*] so that it would not go astray, yet it is now [*insert
number*] days after the date we notified to you. Since we have
neither received a copy of any observations as required by clause
2.10 nor have we received a certificate of practical completion of
the sub-contract works issued by the architect under clause 35.16 of
the main contract, we are concerned.

We should be pleased if you would ask the architect to carry out his
duties under clause 35.16 of the main contract.

Yours faithfully

9.01.6 Letter if contractor's observations are inaccurate
This letter is only suitable for use with NSC/C

To the Contractor
(Copy to the Architect

Dear Sirs

[*Heading*]

We acknowledge receipt of your letter of the [*insert date*] containing a copy of the observations which you have sent to the architect following our notice under clause 2.10 of the sub-contract conditions.

We note that, flying in the face of evidence on site, you consider that practical completion has not taken/will not take [*delete as appropriate*] place in accordance with our notice. We have taken dated photographic records of the situation on site and we refute your observations in full. If the architect does not certify practical completion of the sub-contract works in accordance with our clause 2.10 notice, we shall seek immediate arbitration on the matter.

Yours faithfully

9.01.7 Letter if architect certifies the wrong date as practical completion

This letter is only suitable for use with NSC/C

Registered post/recorded delivery

To the Contractor
(Copy to Architect)

Dear Sirs

[*Heading*]

We have received the architect's certificate of practical completion of the sub-contract works dated [*insert date*]. We do not agree with the date therein certified as the date of practical completion because [*give reasons*].

We should be pleased if you would formally convey our objections to the architect and request him to withdraw the certificate and issue a fresh certificate including the correct date.

If you consider that a meeting would be helpful, we will be happy to attend.

Yours faithfully

9.02 Defects liability

In each form the provisions for defects liability are to be found after the clause dealing with practical completion, thus echoing the main contract arrangement. The provisions are there for your benefit. They give you the right to return to the site after the date of practical completion and rectify defects. This will obviously cost you less than if the contractor employed another sub-contractor to carry out the rectifications and charged the cost to you. The sub-contracts refer to defects liability, not to maintenance, although the clause is often referred to throughout the industry as a maintenance clause.

Your obligation is to make good all defects, shrinkages and other faults. Although this sounds very broad, it is to be interpreted *ejusdem generis* as defects, shrinkages *and other faults like defects and shrinkages*. It is in fact qualified still further because your obligation exists only if the defects and shrinkages are due to materials or workmanship which are not in accordance with the sub-contract, or to frost occurring before the date of practical completion of the sub-contract works.

Therefore, although there may be present what an objective observer may refer to as a defect, you will be obliged to rectify it only if that defectiveness is due to failure to conform to the contract. For example, if you are a plastering sub-contractor whose sub-contract calls for the application of two-coat plastering on block walls, it might be expected that you will not be able to absorb large irregularities in the block wall face in such a thin coat. The finished plaster surface will not be as smooth as if the now comparatively rare three coat plastering system was specified. What would qualify as a defect in respect of one specification, however, would not be the same as a defect under the other specification and your obligation to rectify defects would be qualified accordingly.

You are to make good at your own cost and in accordance with any direction of the contractor. Under the main contract, the architect will prepare a list of defects at the end of the defects liability period and issue them to the main contractor. In practice, the contractor will then split up the list into items which are his responsibility and items which are the responsibility of the various sub-contractors. He will probably do this quite quickly and it may well be that some of the items on your list are not your responsibility. In this case you should notify the contractor immediately (**document 9.02.1**).

If the reason why it is not your responsibility is that the defect is not due to the workmanship or materials not being in accordance with the sub-contract, a different letter is indicated (**document 9.01.2**).

The defects liability clause is expressed to be without prejudice to your obligation to accept a similar liability to that of the contractor under the main contract to remedy defects in the sub-contract works. In DOM/1 and NSC/C the clause is subject to clause 18 (partial possession) of the main contract. DOM/2 is also subject to the partial possession clause of the main contract, which in that instance is clause 17 of CD 81. NAM/SC does not have this proviso because partial possession is not a standard clause in IFC 84. The proviso clearly refers to the situation where defects liability periods start and end at different times for the various parts taken into possession.

Under the main contracts JCT 80, CD 81 and IFC 84 the architect may instruct that a defect is not to be made good, and if this is the case a deduction may be made from the contract sum. Each sub-contract makes provision for the requirement to be stepped down to the sub-contractor and for an appropriate deduction to be made from the sub-contract sum in respect of such defects. Nowhere does it state in the sub-contracts (or main contracts) how an 'appropriate' deduction is to be calculated. The quantity surveyor will have his ideas and no doubt the contractor will have his own. If the deduction or a pro rata share is merely to be passed down from the contractor to you, he is likely to be less interested in its precise calculation. You, of course, will be very interested and you must protest if you consider it is at all unreasonable. Fortunately, the recent case of *William Tomkinson and Sons Ltd* v. *The Parochial Church Council of St Michael and Others* (1990) 6 Const LJ 319 has established the common sense position that an 'appropriate' deduction is what it would have cost you to make good the defect (**document 9.02.3**).

9.02.1   Letter if some defects are not your responsibility

To the Contractor

Dear Sirs

[*Heading*]

Thank you for the schedule of defects sent to us on the [*insert date*].

We have carried out a preliminary inspection and we are making arrangements to make good most of the items on your schedule. However, we do not consider that the following items are our responsibility for the reasons stated:

[*List giving reasons*]

We shall, of course, be happy to attend to such items if you will let us have your written agreement to pay us daywork rates for the work.

Yours faithfully

9.02.2　Letter if defect not due to workmanship or materials not in accordance with the sub-contract

---

To the Contractor

Dear Sirs

[*Heading*]

Thank you for your letter of the [*insert date*] enclosing a schedule of defects which you require us to make good.

We have carried out a preliminary inspection and we would agree that most of the items fall into the category of what a reasonable person would term defective. There are, of course, several reasons why something may be defective; incorrect specification or design for example. The only category which it is our responsibility to make good under the provisions of clause 14.3 [*substitute '15.3' when using NAM/SC or '2.12' when using NSC/C*] is a defect due to workmanship or materials not being in accordance with the sub-contract, or to frost occurring before practical completion of the sub-contract works.

[*continued*]

9.02.2   *contd*

Clearly, the defects on your schedule are not due to frost damage. Equally clearly they are not defects when viewed in the light of the sub-contract requirement. In this connection, we draw your attention to [*give reference number and page of specification, etc.*] from which you can see that we have carried out the work/provided materials [*delete as appropriate*] strictly in accordance with the terms of the sub-contract. We suggest that what is required is an upgrading of the specification/redesign of the items [*delete as appropriate*] and we should be pleased to submit our quotation for the additional work if so invited.

Yours faithfully

9.02.3   Letter if unreasonable deduction made in respect of items not to be made good

To the Contractor

Dear Sirs

[*Heading*]

Thank you for your letter of the [*insert date*] from which we note that, under the provisions of clause 14.4 [*substitute '15.4' when using NAM/SC or '2.13' when using NSC/C*] you instruct us that we are not to make good items [*insert numbers*] on the schedule of defects we received from you yesterday.

You have also notified us that you intend to deduct the sum of [*insert amount*] in respect of the items not made good. This amount is completely unreasonable. It may well reflect the cost of getting the work done by some other firm, but it bears no relationship to what it would cost us to carry out the work. The clause refers to an 'appropriate deduction' and we are advised that there is recent case law which supports this commonsense approach.

[*continued*]

9.02.3 *contd*

We are prepared to agree to the sum of [*insert amount*] being deducted in respect of these items. This is a generous estimate of our costs. If you feel unable to agree to our proposal, we are quite prepared to make good the items on the understanding that no deduction will be made.

If you persist in making this unjustified deduction, we will take further advice immediately, with a view to recovery.

Yours faithfully

Chapter 10

Determination

10.01 Introduction

Under each of the forms of sub-contract under consideration you have the right to determine your employment under certain circumstances. The contractor has similar rights, but of course the circumstances are rather different. Note that it is your employment which is determined, not the contract. It is clear that, after determination, the contract remains in existence to govern the rights of the parties thereafter. The grounds for determination are not generally such as would give either you or the contractor the right to bring the contract to an end at common law. These rights are in addition to your common law rights and this is expressly stated.

Determination is a drastic step which should not be taken except as the last resort. In view of the serious nature of determination, the courts or an arbitrator will look to see that the party carrying out the determination has acted properly in every respect. If the determination is judged to be wrongful, it might well amount to repudiation of the contract. In such a case, the party attempting to determine may be liable for considerable sums in damages.

For this reason, the party determining must be sure of its grounds and follow the procedures to the letter. The law reports abound with instances of sloppily worded notices, certificates and instructions. With this in mind, it is a good idea to incorporate the actual wording of the sub-contract in notices wherever possible so that there is no possible doubt that a notice is being given under a particular clause for a particular purpose.

The sub-contracts also spell out the consequences of determination and the position if the contractor's employment under the main contract is determined.

10.02 Determination by the contractor

Clause 29 in sub-contracts DOM/1 and DOM/2, clause 27 of NAM/SC, and clauses 7.1 to 7.5 of NSC/C, deal with determination of your employment by the contractor. It is expressed to be without prejudice to the contractor's other rights or remedies and the wording in each sub-contract is virtually identical. The grounds for determination are if you default by:

- without reasonable cause, wholly suspending the sub-contract works before completion; or
- without reasonable cause, failing to proceed with the sub-contract works in accordance with the agreed programme details and reasonably in accordance with the progress of the works; or
- refusing or neglecting after written notice from the contractor to remove defective work or improper materials or goods and by your refusal or neglect the works are materially affected; or
- wrongfully failing to rectify defects, shrinkages or other faults in the sub-contract works if the rectification is in accordance with your sub-contract obligations; or
- failing to comply with the provisions forbidding assignment or sub-letting without consent. It is a brave or foolish contractor who attempts to determine on the basis of any of these grounds, unless the circumstances are crystal clear.

Wholly suspending the sub-contract works

The sub-contract works must be totally suspended. If any kind of work is proceeding or if you were simply progressing very slowly, the contractor would not be entitled to determine. Even if you had totally suspended the sub-contract works, you might have done so for a reasonable cause, for example because you were not getting the co-operation you were entitled to receive.

Failing to proceed in accordance with agreed programme details

It has already been noted that the programme details may be very sparse and the programme itself is unlikely to be a contract document (see Chapter 5). Whether you have so failed will be a matter of fact. Your obligation to work reasonably in accordance with the progress of the works is very broad. Again, if you have reasonable cause, such as lack of proper directions, the contractor cannot determine on this ground.

Refusing to remove defective work

Note that the works must be materially affected by your refusal or neglect. This must be the last resort in the kind of circumstances where the works cannot be allowed to proceed until the defective work is rectified. There are other more sensible ways of dealing with failure to comply, for example clause 4.5 of DOM/1.

Failing to rectify defects in accordance with sub-contract obligations

This ground probably looks in particular to your obligations to rectify defects. A significant failure is indicated which precludes the use of less severe remedies.

Failure to comply with provisions regarding assignment or sub-letting

Whether you have attempted to assign your sub-contract or to sub-let any of your obligations under it will be a matter of fact easily demonstrated. Attempting to assign without consent is a serious matter, sub-letting without consent less so.

Under DOM/1, DOM/2 and NAM/SC, the procedure then is that the contractor must send you a notice by registered post or recorded delivery specifying the default. If you do not agree with the notice for whatever reason you must write to the contractor immediately (**document 10.02.1**). If the default notice is justly served, you must write to the contractor to record the fact that you have discontinued the default (**document 10.02.2**). If you continue with the default for ten days after you have received the notice, the contractor may forthwith determine your employment by written notice. In practice, it is not easy for the contractor to know when you have received the notice and he will usually play safe by giving you an extra day. If you disagree with the determination notice, it may be worth writing to the contractor before you seek arbitration (**document 10.02.3**).

The contractor must act within ten days after he is first entitled to determine, so the last date for him to send you notice of determination is twenty days after you receive his initial default notice. Presumably, if the contractor fails to act within the time limit, he would be obliged to start the whole procedure again because a determination notice served outside the 10 day limit would be invalid.

If you stop the default, but default again in the same way later, the contractor is not obliged to serve another default notice, he can simply serve notice of determination. The contractor must be sure that your default is the same as before or he runs the risk of having repudiated the contract. If you think he has made this mistake, you can either obtain expert advice with a view to immediate arbitration or you can ask him to withdraw his notice (**document 10.02.4**).

The grounds in the equivalent clause of NSC/C are identical, but the procedure is different to reflect the architect's involvement. The first step is for the contractor to inform the architect of the default, together with any written observation you may have. This suggests, although the clause does not expressly say so, that the contractor will have warned you of his intention or at least he will have written to you about the default in order to ascertain your views. If he does write to you, you will, of course, recite the facts as you know them together with any evidence which supports your contentions (**document 10.02.5**).

When the architect receives the notice from the contractor together with your comments, he may decide not to instruct any further action, but it is possible that he will instruct the contractor to send you a default notice. **Document 10.02.1** is suitable if the notice is not justified and **document 10.02.2** if it is justified. You have 14 days in which to discontinue your default, failing which the contractor may, within ten days as before, serve notice of determination. The architect may make such determination subject to his further instruction to the contractor under the provisions of clause 35.24.6.1 of the main contract, but of course you would not know that. He is more likely to thus fetter the actions of the contractor if you have made a very strong case in your initial observations about your alleged default. There is a proviso to each of these sub-contracts that the notice must not be given unreasonably or vexatiously. If, under any of the sub-contracts, the contractor's notice of determination is premature in relation to the original default notice, it is invalid and it is probably wise to challenge it immediately (**document 10.02.6**).

If you become insolvent, the contractor may forthwith determine your employment and, under NSC/C, determination is automatic. If you are insolvent, determination may be the least of your worries. The contractor may also determine without prior notice on the grounds of your corruption.

10.02.1   Letter if default notice served
Registered post/recorded delivery

To the Contractor
(Copy to Architect when using NSC/C)

Dear Sirs

[Heading]

We are in receipt of your letter of the *[insert date]* which, apparently, you intend to be a default notice in accordance with clause 29.1 *[substitute '27.1' when using NAM/SC or '7.1' when using NSC/C]*.

[Add either:]

The purported notice contains a serious error.

[Or:]

The substance of the default specified in your purported notice is incorrect.

[Or:]

The alleged default specified in your purported notice is a direct result of your default *[explain]*.

[continued]

10.02.1 *contd*

[*Or:*]

You have failed to specify the alleged default.

[*Then add:*]

Therefore, we do not consider that your letter constitutes a valid notice of default under the sub-contract. Any attempt to determine our employment, which relies on your letter, will be strongly resisted. It might well amount to a repudiation of the sub-contract for which substantial damages will be payable.

Yours faithfully

10.02.2   Letter if default notice served justly
Registered post/recorded delivery

To the Contractor
(Copy to Architect when using NSC/C)

Dear Sirs

[Heading]

We are in receipt of your letter of the *[insert date]* which you
have sent as a default notice in accordance with clause 29.1
[substitute '27.1' when using NAM/SC or '7.1' when using NSC/C]
of the conditions of sub-contract.

We do not consider that the circumstances merited such a notice,
which does not contribute to good relations. However, we are
pleased to be able to inform you that *[insert the steps being taken
to remove the default]*.

Yours faithfully

10.02.3 Letter if notice of determination served
Registered post/recorded delivery

To the Contractor
(Copy to Architect when using NSC/C)

Dear Sirs

[*Heading*]

We are in receipt of your letter of the [*insert date*] which purports to determine our employment under clause 29.1 [*substitute '27.1' when using NAM/SC or '7.1' when using NSC/C*] of the sub-contract.

[*Add either:*]

We refer you to our letter of the [*insert date*] from which you will see that we contend that your purported default notice is invalid.

[*Or:*]

If we assume that your original default notice is valid, which we deny, your purported determination notice was not given within 10 days of the alleged continuance of the default.

[*continued*]

10.02.3   *contd*

[*Or:*]

Such determination is not valid because [*insert reasons*].

[*Then:*]

We are, therefore, proceeding with the sub-contract works.

Yours faithfully

10.02.4   Letter if default is not a repeated default
Registered post/recorded delivery

To the Contractor
(Copy to Architect when using NSC/C)

Dear Sirs

[*Heading*]

We are in receipt of your letter of the [*insert date*] which purports
to determine our employment under clause 29.2 [*substitute '27.1'*
when using NAM/SC or '7.1' when using NSC/C] of the
sub-contract. We note that you rely upon our alleged repetition of
an alleged default for which you gave us a notice dated [*insert*
date].

We deny that we were ever in default as mentioned in your notice.
Moreover, the alleged default on which you now rely to support
your attempted determination is not the same alleged default
mentioned in such notice and, therefore, your reliance is misplaced.

Your purported determination is, therefore, invalid and we are
entitled to treat it as a wrongful repudiation of the contract on your
part. We reserve all our rights and remedies until we have taken
further advice on this matter. Meanwhile, and without prejudice to
any action we may deem it necessary to take, we intend to proceed
with the sub-contract works.

Yours faithfully

10.02.5   Letter if contractor invites observation on default
This letter is only suitable for use with NSC/C

To the Contractor

Dear Sirs

[*Heading*]

Thank you for your letter of the [*insert date*] from which we note that you intend to inform the architect that we are in default as specified in clause 7.1.1/.2/.3/.4 [*delete as appropriate*] of the sub-contract.

Under the provisions of clause 7.1 you are obliged to send to the architect any written observations in regard to the default. Our observations are as follows:

[*State why you are not in default or if you are in default, make it clear that you have or are about to stop the default then, if appropriate, add:*]
Please also notify the architect that if you proceed to issue a default notice we shall treat it as a threat on your part to repudiate the sub-contract.

Yours faithfully

10.02.6 Letter if determination is premature
Registered post/recorded delivery

To the Contractor
(Copy to Architect when using NSC/C)

Dear Sirs

[*Heading*]

We are in receipt of your notice dated [*insert date*] which purports to determine our employment under clause 29.1 [*substitute '27.1' when using NAM/SC or '7.1' when using NSC/C*] of the sub-contract.

Your original notice was received on the [*insert date*]. The Post Office will be able to confirm that date. Your notice of determination was, therefore, premature and of no effect. It may amount to a repudiation of the sub-contract for which we can claim substantial damages. Without prejudice to any of our rights and remedies in this matter we have corrected the default specified in your notice and we will proceed with the sub-contract works while we take advice on this matter.

Yours faithfully

10.03   Consequences of determination by the contractor

The consequences are dealt with by clauses 30.3 and 30.4 of DOM/1 and DOM/2, 27.3 of NAM/SC, and 7.4 and 7.5 of NSC/C. The situation is very similar in each case. If the contractor determines your employment the sub-contract provides that he or another sub-contractor can use your plant and equipment to finish the sub-contract works. If required you must assign the benefit of any agreements to the contractor without payment, he may pay your suppliers or sub-sub-contractors direct and you must remove all your temporary buildings, plant and equipment from the site if the contractor so requests. Moreover, you are not entitled to receive any further payment until after the completion of the sub-contract works.

Each sub-contract has differences. Under DOM/1 and DOM/2, after completion of the sub-contract works, you 'may apply' to the contractor and he 'shall' pay the value of work done and materials supplied, provided that the value has previously been included in interim payments. The wording is curious. It may mean that the contractor is bound to carry out the payment procedure or it may mean that he is only obliged to pay if you apply. In order to avoid any possibility of doubt, it is suggested that you should always apply (**document 10.03.1**). The contractor may deduct not only cash discount but also the amount of any direct loss and/or damage caused to him by the determination. When he has carried out that exercise, it is likely that the contractor will be looking to you for extra money.

NAM/SC expresses the position in a different way, but it probably amounts to the same thing when interpretation is given to loss and/or damage. There is no requirement for you to apply before you are paid, but the contractor is to draw up an account showing the amount of direct loss and/or damage he has suffered due to the determination, and the amount he has paid to you before determination, and if the total is less than the sub-contract sum as adjusted, the difference is a debt payable to you by the contractor. If the total is more, the difference is a debt payable to the contractor. If you consider that money is due, you should always remind the contractor on completion of the sub-contract works (**document 10.03.2**).

An important clause is 27.3.3, in which you undertake not to contend that the contractor has suffered no loss if he tries to recover money from you after determination, as he is obliged to do under clause 3.3.6(b) of the main contract. The fact is that the contractor has suffered no loss under that clause and will not do so if he fails to recover the money from you because he is liable to pay the employer only such money as he has managed to recover. The clause is there because in any proceedings he would normally have to prove his loss before recovering damages and the rule is that if there is no loss there can be no damages. It is not certain that the clause is effective in solving that problem and if the contractor writes to you for his money, you should try **document 10.03.3**.

Under NSC/C, it is for you to decide whether to apply to the contractor after completion of the sub-contract works (**document 10.03.4**) If you do apply, the contractor must pass your application to the architect who must ascertain the expense and direct loss and/or damage incurred by the employer as a result of the determination. In paying any certificate then issued by the architect in respect of sub-contract work not previously paid, the employer may deduct such expense, loss and/or damage. In turn, the contractor may deduct cash discount and any direct loss and/or damage caused him by the determination.

**10.03.1	Letter applying for payment after determination**
This letter is only suitable for use with DOM/1 or DOM/2

To the Contractor

Dear Sirs

[*Heading*]

We note that the sub-contract works are now complete under clause
29.3.1 of the sub-contract.   The matter of your alleged
determination of our employment under the sub-contract is the
subject of a dispute between us.   Without prejudice to such dispute
or any of our rights and remedies, we hereby apply, pursuant to
clause 29.4 of the sub-contract, for payment of the value of work
executed and goods and materials supplied by us under the
sub-contract.

Yours faithfully

10.03.2   Letter applying for payment after determination
This letter is only suitable for use with NAM/SC

To the Contractor

Dear Sirs

[*Heading*]

We note that the sub-contract works are complete. In accordance with clause 27.4 of the sub-contract, you are obliged to set out an account as detailed in clause 27.4 and pay us the money shown as owing. This notice is without prejudice to any of our rights and remedies in relation to the alleged determination of our employment.

Yours faithfully

10.03.3   Letter if contractor tries to recover money under clause 27.3.3
This letter is only suitable for use with NAM/SC

To the Contractor

Dear Sirs

[*Heading*]

We are in receipt of your letter of the [*insert date*] from which we note that you are requesting payment of money which you state it is your duty to recover from us under clause 3.6(b) of the main contract. The matter of the alleged determination is the subject of a dispute between us.

We also note that under the provisions of clause 27.3.3 of the sub-contract we are not to contend that you have suffered no loss. We make no such contention. But if you take any action to secure payment, we give notice that we shall require you to present evidence of the amount of the loss you have suffered.

Yours faithfully

10.03.4 Letter applying for payment after determination
This letter is only suitable for use with NSC/C

To the Contractor

Dear Sirs

[*Heading*]

We note that the sub-contract works have been completed as referred to in clause 7.4.1 of the sub-contract.

The alleged determination of our employment is the subject of dispute between us. Without prejudice to the result of such dispute or to any rights or remedies which we may possess, we hereby apply under the provisions of clause 7.5 and require you to carry out your obligation to pass our application to the architect so that he can issue a certificate for the value of work executed and goods and materials supplied by us for the sub-contract works.

Yours faithfully

10.04   Determination by the sub-contractor

The determination procedure is covered by clause 30 of DOM/1 and DOM/2, clause 28 of NAM/SC, and clauses 7.6 and 7.7 of NSC/C. Your power to determine is said to be without prejudice to any other rights and remedies you may possess. The grounds for determination under each sub-contract are defaults by the contractor. An important proviso states that there must not be a remedy under any other provision of the sub-contract which would adequately recompense you. If there is, you must use that remedy and you may not determine your employment. Such a remedy might be a claim for loss and/or expense or suspension for failure to pay.

The grounds are if the contractor:

- wholly suspends the works before completion without reasonable cause; or
- without reasonable cause, fails to proceed with the works so that the reasonable progress of the sub-contract works are seriously affected; or
- fails to pay in accordance with the sub-contract (not NSC/C).

The final ground is missing from NSC/C because under this form you are already adequately protected by the direct payment provisions if the contractor defaults.

You must be absolutely certain of your ground before you put the machinery into motion.

Wholly suspends the works

Total suspension of the works may appear straightforward and easy to identify, but whether the contractor has a reasonable cause is always open to dispute. The same kind of considerations apply to this ground as apply to you wholly suspending the sub-contract works when the contractor is considering determination. An important difference which you must not overlook is that here the contractor has to have wholly suspended the main contract works including the sub-contract works. This ground is not adequate to allow you to determine if he simply suspends the sub-contract works alone. In that instance, the next ground is appropriate.

Failure to proceed with the works so that reasonable progress of the sub-contract works is seriously affected

Failure to proceed so that your reasonable progress is seriously affected is less stringent than the first ground, but it is difficult to identify with certainty. Quite apart from whether the contractor has reasonable cause, you have to decide at what point your reasonable progress is seriously affected. Is there a difference between seriously and noticeably affected?

Fails to pay in accordance with the sub-contract

You may not simply determine because you have not received the amount in your application; the contractor may be able to argue that your application does not exclusively refer to work properly carried out. He is under no obligation to pay for defective work. It is probably safest to rely on total failure to pay.

The procedure is similar to determination by the contractor. If you are satisfied that the contractor has indeed defaulted in one of the ways specified, you may serve him with a notice by registered post or recorded delivery specifying the default (**document 10.04.1**). If he continues his default for 10 days you may thereupon serve written notice by registered post or recorded delivery determining your employment. You may use **document 10.04.2**. If he stops the default, but at some time in the future he defaults *in the same way*, you may determine without having to serve the intial default notice again. There are two differences under NSC/C. You must send the architect a copy of the default notice and the contractor has 14 days in which to discontinue his default.

Under DOM/1, DOM/2 and NAM/SC, there is an important clause which stipulates that if you have suspended the sub-contract works under clause 21.6 or 19.6 respectively (failure to pay), you may not issue a notice of determination until 10 days after the date of commencement of the suspension. Therefore, if you serve a default notice and two days later suspension of the works commences, it will be the date of commencement of suspension and not the date of receipt of the default notice which decides when you may give notice of determination.

10.04.1   Notice of default before determination
Registered post/recorded delivery

To the Contractor
[Copy to Architect when using NSC/C]

Dear Sirs

[Heading]

We hereby give you notice under clause 30.1.1/.2/.3 *[delete as appropriate or substitute '28.1.1/.2/.3' as appropriate when using NAM/SC or '7.6.1/.2' as appropriate when using NSC/C]* of the conditions of sub-contract that you are in default in the following respect:
[Insert details of the default with dates if appropriate]

If you continue the default for 10 *[substitute '14' when using NSC/C]* days after receipt of this notice, we may thereupon determine our employment under this sub-contract without further notice.

Yours faithfully

10.04.2   Letter determining employment
Registered post/recorded delivery

To the Contractor
(Copy to Architect when using NSC/C)

Dear Sirs

[*Heading*]

We refer to our notice dated [*insert date*] and received by you, as notified to us by the Post Office, on [*insert date*].

We hereby give notice in accordance with clause 30.1 [*substitute '28.1' when using NAM/SC or '7.6' when using NSC/C*] of the sub-contract that we forthwith determine our employment under this sub-contract without prejudice to any other rights and remedies which we may possess.

We are making arrangements to move all our temporary buildings, plant, etc, and materials from the works and we will write to you again within the next week regarding our financial entitlement.

Yours faithfully

10.05   Consequences of determination by the sub-contractor

After you have given notice of determination the procedure is governed by clause 30.2 under DOM/1 and DOM/2, clause 28.3 under NAM/SC, and clause 7.7 under NSC/C.

You must remove all your plant and equipment from site as soon as you reasonably can. In so doing, you must take such precautions as will prevent injury, death or damage of the classes in respect of which you are obliged to indemnify the contractor under the insurance clauses. You are entitled to be paid the following:

- The total value of work completed at the time of determination.
- The total value, as if it was a variation, of work begun but not completed.
- Any sum ascertained in respect of loss and/or expense whether ascertained before or after determination.
- The cost of materials or goods for which you have paid or for which you are legally bound to pay.
- The reasonable cost of removal.
- Any direct loss and/or damage which you have incurred as a result of the determination (not DOM/1 or DOM/2).

Although DOM/1 and DOM/2 make no specific provision for you to be paid direct loss and/or damage caused by the determination, your common law rights are available for you to recover appropriate damages. If the contractor agrees, your claim can be dealt with at the same time as your other payments. It will be less costly than litigation or arbitration for both parties. It is worthwhile asking the contractor if he wishes to deal with it informally (**document 10.05.1**).

The contractor is unlikely to be in a hurry to carry out the appropriate calculations in order to pay you promptly. There is no express stipulation regarding the time for such payment, but probably it must be within a reasonable time. Since you are the injured party, you should not have to wait for your money for as long a period as when the contractor determines. You are clearly entitled to payment before the completion of the sub-contract works. If the contractor drags his heels, you must send him a letter (**document 10.05.2**). Unfortunately, the most likely consequence of determination is probably arbitration.

10.06   Determination of the contractor's employment under the main contract

The basic rule is that your employment determines if the contractor's employment is determined under the main contract. The consequences depend upon the reason for determination. Under DOM/1 and

DOM/2, the consequences are as if you had determined your own employment under clause 30. Under NAM/SC, the consequences vary. If determination is for so-called 'neutral causes' under clause 7.8 of the main contract, the procedure is according to clause 28.3 except that you are not entitled to loss and/or damage. NSC/C adopts a similar approach under clauses 7.8 and 7.9.

10.05.1   Letter asking the contractor to deal with a common law claim after determination

This letter is only suitable for use with DOM/1 or DOM/2

WITHOUT PREJUDICE

To the Contractor

Dear Sirs

[*Heading*]

We refer to our notice of determination dated [*insert date*].

These circumstances entitle us to a claim for damages against you. Clause 30.2 of the conditions of sub-contract sets out the rights and duties of the parties following determination, but it does not appear to deal with direct loss and/or damage caused to us by the determination. It seems, therefore, that we must recover the damages by way of arbitration or litigation and our legal advisors are currently considering the matter.

If you are prepared to meet us to discuss our claim with a view to reaching a reasonable settlement, we will take no immediate legal proceedings.

Please inform us by [*insert date*] if you agree to this suggestion and let us know a convenient date.

Yours faithfully

10.05.2   Letter if contractor is slow in paying after determination
Registered post/recorded delivery

To the Contractor

Dear Sirs

[*Heading*]

We refer to our determination notice dated [*insert date*].

Clause 30.2 [*substitute '28.3.2' when using NAM/SC or '7.7' when using NSC/C*] obliges you to pay us in accordance with the rules set out therein. On [*insert date*], we sent you our detailed calculation of the amount due; a further copy is enclosed herewith. You have failed to take any action and our telephone calls to your office have met with prevarication and obfuscation.

We hereby give notice that you are in breach of contract and if we do not receive a substantial payment on account by [*insert date*] we will take immediate action for recovery.

Yours faithfully

Chapter 11

Disputes and their Resolution

11.01 Introduction

The resolution of disputes under the standard forms of sub-contract tends to be more complex than resolving disputes under the main contract equivalent. That is because the sub-contractor is potentially looking at one or both of the contractor and the employer for a remedy. Many decisions made by the architect affect the sub-contractor and this is particularly the case where NSC/C is concerned and the architect is called upon to issue certain certificates and take other actions which directly concern the nominated sub-contractor. If the sub-contractor wishes to pursue a remedy against the employer under that sub-contract, he is obliged to use the name-borrowing facilities, i.e. to bring an arbitration against the employer in the contractor's name. This is because there is no direct contractual relationship between the sub-contractor and the employer. Sub-contracts also make use of adjudication specifically to deal with set-off. Set-off and adjudication are dealt with in section 6.05 and 6.06.

It is always preferable to settle disputes by negotiation and various methods of attaining this are being explored under the general heading of Alternative Dispute Resolution (ADR). None of the standard forms provide for this process yet. It can be formally written into the terms, but the drafting requires expert guidance if the provision is to be more than an excuse for one party to delay resolution of the dispute.

11.02 Arbitration between sub-contractor and contractor

DOM/1 and DOM/2 deal with arbitration in clause 38 and NAM/SC and NSC/C deal with arbitration in clauses 35 and 9 respectively. They are very much in common form. Importantly, arbitrations are to be conducted in accordance with the JCT Arbitration Rules 1988 which, properly used, offer the parties several advantages not otherwise available. The Rules provide for three basic procedures, one of which should suit the issues in dispute and the desires of the parties in relation to the issues. The procedures are:

● *Procedure without a hearing*: Designed for a situation where oral evidence is not necessary. It is operated on the basis of written submissions and an inspection of the site by the arbitrator if appropriate. This is intended to be the normal procedure.
● *Short procedure with a hearing:* This procedure is ideally suited to disputes regarding the quality of workmanship, compliance with a specification and the like. The short hearing can take place on site with the arbitrator giving an immediate award in some instances. The parties must agree before this procedure is used.
● *Full procedure with a hearing:* This is what has regrettably become the norm for arbitration hearings – a complete procedure which is almost identical to a full court action. This is only really necessary when oral evidence is required.

Some sub-contractors make a practice of threatening arbitration whenever any kind of dispute arises and long before the parties have adopted intransigent positions. In such instances, it is used as a bargaining tool. That is a very dangerous and shortsighted policy. Even more shortsighted is to issue a notice of arbitration before properly considering all avenues of resolution. Arbitrations are relatively easy to start, but difficult to stop unless both parties agree.

If you are satisfied that there is no other course of action, you should start the procedure by serving notice on the contractor (**document 11.02.1**). The purpose of the notice is to appoint an arbitrator. If the parties do not concur in the appointment, the appointing body named in the contract may be asked to make the appointment. The appointing body is to be inserted in the Appendix part 14 (DOM/1 and DOM/2), or in NAM/T, section 1, item 18 (NAM/SC), or in NSC/T, part 3, item 7 (NSC/C). When you apply, the appointing body will send you details of its procedure and a form to be completed with the relevant details.

It is obviously better for the parties to agree an arbitrator, because they have control over the appointment. Once an appointing body has been asked to appoint, the parties have lost control and they are obliged by law to accept whoever is appointed. Lists of arbitrators are kept by the Chartered Institute of Arbitrators who will be pleased to assist with names of suitable persons. The address is: The Secretary, Chartered Institute of Arbitrators, International Arbitration Centre, 24 Angel Gate, City Road, London EC1V 2RS.

If you receive a notice to concur in the appointment of an

arbitrator, you should immediately seek expert advice. **Document 11.02.2** is a suitable letter to send to the contractor while you are seeking such advice.

It would be inappropriate to suggest further typical letters, because you should not attempt to conduct arbitration proceedings yourself. Both authors have encountered situations where one party has been astounded to find that the Preliminary Meeting, which he imagined to be an informal general discussion between the parties, is in fact a formal arena within which binding dates and procedures are established for the conduct of the arbitration.

The remainder of the arbitration clause deals with the possibility of joining disputes about the same issues under main and sub-contracts, the powers of the arbitrator, when arbitration may commence, agreement as to appeals, and arrangements if the arbitrator ceases to act for any reason before making his final award.

11.02.1 Letter requesting concurrence in the appointment of an arbitrator

Registered post/recorded delivery

To the Contractor

Dear Sirs

[*Heading*]

We hereby give you notice that we require the undermentioned dispute or difference between us to be referred to arbitration in accordance with article 3 [*substitute '4' when using NAM/SC or NSC/C*] of the sub-contract between us dated [*insert date*]. Please treat this as a request to concur in the appointment of an arbitrator under clause 38.1 [*Substitute '35.1' when using NAM/SC or '9.1' when using NSC/C*].

The dispute or difference is [*insert brief description*].

We propose the following three persons for your consideration and require your concurrence in the appointment within 14 days of the date of this notice, failing which we shall apply to the President or Vice-President of [*insert name of the appointing body*].

The names and addresses of the persons we propose are: [*list names and addresses of three persons*]

Yours faithfully

11.02.2   Letter, if sub-contractor receives notice to concur
Registered post/recorded delivery

To the Contractor

Dear Sirs

[*Heading*]

We are in receipt of your letter dated [*insert date*] by which you refer to certain matters and ask us to concur in the appointment of an arbitrator.

Your letter has been passed to our legal advisors and we have instructed them to reply to you on our behalf.

Yours faithfully

11.03   Name borrowing

Only NSC/C provides for 'name borrowing'. That is the name usually given to the procedure whereby the sub-contractor, as a contractual right, may require the contractor to allow him to use the contractor's name in arbitration proceedings against the employer. The procedure is included in NSC/C to reflect the involvement of the employer (through his architect) in the nomination of sub-contractors. It enables the sub-contractor to pursue certain disputes directly with the employer. They are:

- Under clause 2.7, if the sub-contractor is aggrieved by the architect's failure to give his consent as required by clause 2.3 (extension of time), or by the architect's failure to give his consent within the specified time period, or by the terms of the consent.
- Under clause 3.11, if the sub-contractor wishes to challenge the architect's specified clause empowering an instruction.
- Under clause 4.20, if the sub-contractor is aggrieved by the amount certified by the architect or the architect's failure to certify.

Document 11.03.1 is a suitable letter to the contractor. It must be remembered that only the issues noted above may be the subject of name borrowing.

A pre-requisite is that the sub-contractor must give the contractor such indemnity and security as the contractor may reasonably require. Because the proceedings will be in the contractor's name, the contractor will be liable for any costs resulting from the proceedings. If, for example, the employer put forward a successful counterclaim, the contractor might face a very substantial bill at the end. If you receive a request for what you believe to be unreasonable security or indemnity, try **document 11.03.2**. The contractor is obliged to allow you to use his name and, if necessary (that is, if necessary in the circumstances) he must also join with you in the arbitration. Usually, however, you will conduct the arbitration directly with the employer and the contractor will take no active part (he might wish to be present as an observer and he would be prudent to do so). If he tries to refuse permission to use his name, make the position clear (**document 11.03.3**).

Before embarking upon this exercise, it is important to obtain good legal advice and all the previous comments relating to arbitration are applicable.

A clause which is sometimes confused with name borrowing, but which is quite separate in its operation and effect, is the clause entitled 'Benefits under Main Contract'. DOM/1 and DOM/2 deal with this topic under clause 22, NAM/SC and NSC/C deal with it under clauses 20 and 1.13 respectively. The provisions of DOM/1, DOM/2 and NAM/SC are identical and provide as follows:

'The Contractor will so far as he lawfully can at the request and cost, if any, of the Sub-Contractor obtain for him any rights or benefits of the Main Contract so far as the same are applicable to the Sub-Contract Works but not further or otherwise.'

NSC/C further provides that any action which the contractor takes in complying with any request of the sub-contractor will be at the sub-contractor's cost and may include the provision of such indemnity and security as the contractor reasonably requires. It, therefore, appears that under the other forms of sub-contract, the contractor is not entitled to ask for indemnity and security for his costs as a pre-condition for action on his part and you should resist suggestions to the contrary (**document 11.03.4**).

If the sub-contractor requests action under these clauses (**Document 11.03.5**), the contractor may be obliged to seek arbitration with the employer to recover benefits which are essentially benefits for the sub-contractor. Unlike the position where name borrowing is concerned, the arbitration would be conducted by the contractor. If the contractor refuses to take action, **document 11.03.6** is suitable.

11.03.1 Letter if the sub-contractor wishes to use the contractor's name

This letter is only suitable for use with NSC/C
Registered post/recorded delivery

To the Contractor

Dear Sirs

[Heading]

Take this as formal notice that under the provisions of clause 2.7/3.11/4.20 *[delete as appropriate]* of the conditions of sub-contract we wish to use your name and if necessary we require you to join with us in arbitration proceedings to decide the following matter: *[briefly describe the matter in dispute]*

Yours faithfully

11.03.2   Letter if contractor requires unreasonable indemnity or security

This letter is only suitable for use with NSC/C

To the Contractor

Dear Sirs

[*Heading*]

We are in receipt of your letter dated [*insert date*] from which we note that you require [*insert the indemnity and security required, using the contractor's own words*]. We refer you to clause 2.7/3.11/4.20 [*delete as appropriate*] of the conditions of sub-contract which refers to your reasonable requirement for indemnity or security. The requirements you set out are not reasonable on any interpretation of those words and we propose the following: [*set out your own reasonable proposals*].

We look forward to receiving your agreement within seven days from the date of this letter. In the absence of such agreement, we shall be obliged to seek legal advice on the matter.

Yours faithfully

11.03.3   Letter if contractor refuses permission to use his name
This letter is only suitable for use with NSC/C
By fax and recorded delivery

To the Contractor

Dear Sirs

[*Heading*]

We are in receipt of your letter dated [*insert date*] and note that you purport to refuse permission for us to use your name in arbitration proceedings.

If you will carefully read clause 2.7/3.11/4.20 [*delete as appropriate*] you will see that it is mandatory upon you to allow us to use your name subject only to providing you with such indemnity or security as you reasonably require. We are and always have been prepared to comply with your reasonable requirements and we should be pleased if you will acknowledge our legitimate entitlement by return of post. Failure to do so will oblige us to examine our legal remedies.

Yours faithfully

11.03.4   Letter if contractor seeks indemnity and security before obtaining benefits for sub-contractor

This letter is not suitable for use with NSC/C

To the Contractor

Dear Sirs

[Heading]

We are in receipt of your letter dated *[insert date]* seeking indemnity and security for your costs before putting into effect your duties under clause 22 *[substitute '20' when using NAM/SC]* of the conditions of sub-contract.

If you will examine the clause in detail, you will find that it makes no provision for us to provide such indemnity or security although it is a very short clause and could very easily have included such a provision as does clause 1.13 of sub-contract form NSC/C. We should be pleased if you would comply with your very clearly defined obligations forthwith.

Yours faithfully

11.03.5 **Letter if sub-contractor requires the contractor to obtain benefits**

To the Contractor

Dear Sirs

[*Heading*]

Under the provisions of clause 22 [*substitute '20' when using NAM/SC or '1.13' when using NSC/C*] of the conditions of contract, we request you to obtain for us the following benefits under the main contract so far as they are applicable to the sub-contract works: [*carefully specify the required benefit*]. We understand that any costs will be our responsibility and we, therefore, further require that you notify us and obtain our written approval on each occasion before incurring any cost.

Yours faithfully

11.03.6   Letter if the contractor refuses to obtain benefits under the main contract

By fax and recorded delivery

To the Contractor

Dear Sirs

[Heading]

We are in receipt of your letter dated *[insert date]* from which it seems that you are not prepared to take action under the contract to obtain for us the benefits of the main contract so far as they are applicable to the sub-contract works. We gave you our request under clause 22 *[substitute '20' when using NAM/SC or '1.13' when using NSC/C]* on the *[insert date]*. Under the terms of the contract, you are obliged to obtain such benefits on our request and there is no provision for you to object. If you do not withdraw your letter and indicate in writing by return of post your willingness to act in accordance with the contract, we shall take positive steps to secure our position including whatever legal remedies are appropriate.

Yours faithfully

**11.04 Arbitration between sub-contractor and employer
under NSC/W**

Under the Employer/Nominated Sub-contractor Agreement NSC/W,
there is provision for arbitration between the parties under clause 11.1.
It should be considered relevant to disputes under or in connection
with NSC/W only. Whether it could be used in a wider application to
resolve any dispute between the employer and the sub-contractor in
connection with the sub-contract is problematical. The arbitration
clause is broadly similar to the arbitration clauses in the sub-contract
forms, and the previous remarks in this chapter apply to this clause
also.

Index